Supermaterials

NATO Science Series

A Series presenting the results of scientific meetings supported under the NATO Science Programme.

The Series is published by IOS Press, Amsterdam, and Kluwer Academic Publishers in conjunction with the NATO Scientific Affairs Division

Sub-Series

I. **Life and Behavioural Sciences**	IOS Press
II. **Mathematics, Physics and Chemistry**	Kluwer Academic Publishers
III. **Computer and Systems Science**	IOS Press
IV. **Earth and Environmental Sciences**	Kluwer Academic Publishers

The NATO Science Series continues the series of books published formerly as the NATO ASI Series.

The NATO Science Programme offers support for collaboration in civil science between scientists of countries of the Euro-Atlantic Partnership Council. The types of scientific meeting generally supported are "Advanced Study Institutes" and "Advanced Research Workshops", and the NATO Science Series collects together the results of these meetings. The meetings are co-organized bij scientists from NATO countries and scientists from NATO's Partner countries – countries of the CIS and Central and Eastern Europe.

Advanced Study Institutes are high-level tutorial courses offering in-depth study of latest advances in a field.
Advanced Research Workshops are expert meetings aimed at critical assessment of a field, and identification of directions for future action.

As a consequence of the restructuring of the NATO Science Programme in 1999, the NATO Science Series was re-organized to the four sub-series noted above. Please consult the following web sites for information on previous volumes published in the Series.

http://www.nato.int/science
http://www.wkap.nl
http://www.iospress.nl
http://www.wtv-books.de/nato-pco.htm

Supermaterials

edited by

Rudi Cloots
Marcel Ausloos
University of Liege,
Liege, Belgium

Marek Pekala
University of Warsaw,
Warsaw, Poland

Alan J. Hurd
Sandia National Laboratories,
Albuquerque, NM, U.S.A.

and

Gilbert Vacquier
University of Provence,
Marseille, Cedex, France

Kluwer Academic Publishers

Dordrecht / Boston / London

Published in cooperation with NATO Scientific Affairs Division

Proceedings of the NATO Advanced Research Workshop on
Supermaterials
Giens, Hyeres, France
September 19–23, 1999

A C.I.P. Catalogue record for this book is available from the Library of Congress.

ISBN 0-7923-6808-8 (HB)
ISBN 0-7923-6809-6 (PB)

Published by Kluwer Academic Publishers,
P.O. Box 17, 3300 AA Dordrecht, The Netherlands.

Sold and distributed in North, Central and South America
by Kluwer Academic Publishers,
101 Philip Drive, Norwell, MA 02061, U.S.A.

In all other countries, sold and distributed
by Kluwer Academic Publishers,
P.O. Box 322, 3300 AH Dordrecht, The Netherlands.

Printed on acid-free paper

Printed in the Netherlands.

TABLE OF CONTENTS

PREFACE

What is a *Supermaterial*? On rare occasions, the coining of a new term brings new ideas to a field by virtue of a shift in viewpoint. A recent example is "complexity", which is a set of phenomena connected together by a core of deep results in mathematics and hydrodynamics; the result is a clan of fields whose kinship had been revealed by the new term itself. Often such appellations merely follow fashionable trends. So it took courage on the part of the launchers of SUPERMAT to promote some vision. In so doing we have tried to show a way to resonate as a new chord in materials science. The first conference on "supermaterials", SUPERMAT and its companion SMART 99 (Proceedings to be found in Superconductor Science and Technology) have begun to reveal new branches of research from the established pathways explored in superconductivity.

At the conclusion of SUPERMAT, held on the Giens Peninsula, Hyères, France, on Sept. 19-23, 1999, participants were as energized by the exciting science that had been covered during the week as they were impressed by the French food, wine, scenery and hospitality. If the definition of a supermaterial is not obvious now after the conference, it was certainly not obvious before it. Nevertheless, the conferees were drawn inexorably together by the conference theme. Perhaps it is not appropriate to try to understand this self organization, which surely is even more complex than the self organization of electrons in a high temperature superconductor, the theory for which still eludes us after 13 years!

A clue to the working definition of a supermaterial can be derived empirically from the topics we discussed at SUPERMAT and SMART 99. In addition to **super**conductors, we heard about magnetic materials of many kinds, including "giant" and even "colossal" ones that presumably trump the performance–if not the inherent interest–of superionic conductors, organic conductors, photoconductors, and even four-hundred-year-old Japanese ceramics. A useful way to look at these topics is through the materials science tetrahedron. We see that processing is a prominent pursuit in superconductor subset. Characterization and theory talks broke new ground in pursuit of e.g. useful optoelectronic phenomena. In synthesis, the parade of new materials combinations, often in groups of four elements or more, was a continuing source of surprise. But a clearly distinctive aspect of the conference that would not have been possible until recently is the wealth of applications reported. Several of them are not included in the following proceedings for not uselessly duplicating reports. We emphasize in bold the contributors to this volume, and mention all oral reports. A poster session was held as well.

From circuits to sensors, supermaterials are making an impact on society. P. Seidel (Jena) made clear the path for supermaterials to impact medical technology, and many others discussed energy applications from fuel cells to current limiter components. In a discussion that focused keenly on one of the limiting factors of Moore's law, Michel **Houssa** (IMEC, Leuven) presented research into alternative gate materials for microcircuits. Currently, silicon oxide gates are approaching a few nanometers (!) and will clearly fail to meet reliability requirements. Alternative oxides and nitrides with the required high dielectric constants were eloquently described.

For the "super"conductors, D. Dew-Hughes (Oxford) amazed us with a discussion of new induction motors based on superconducting, hollow cylinders that self-center in a stator field. With appropriate vanes, the motors become compact pumps, promising for cryofluids and microfluidic devices. Other microdevices were presented by J.-Cl.Villegier (Grenoble) and A.J. Kreisler (SUPELEC, Gif sur Yvette), who discussed remarkable superconducting field effect transistors, bolometers, and IR detectors based on supermaterials. The microwave devices were discussed by E. Silva (Roma3). Also D. Robbes (CRISMAT, Caen) discussed that very low temperature TES sensors could benefit from phonon confinement in YBCO/STO bilayers, while A. Plecenik (Bratislava) discussed YBCO/Metal junctions interface properties from the order parameter symmetry point of view. Much more should be thought about such a "parameter" for future devices, - and from the fundamental point of view claimed Alexander M. **Gabovich** (Kiev). See also the comment by Alexey V. **Nikulov** (Chernogolovka).

A tour de force of high resolution microscopy and theoretical analysis was shown by D.Welch (Brookhaven) in his discussion of the importance of twinning in superconductors. The fundamental understanding of these microstructural aspects of supermaterials may soon fall to the elegant computational techniques covered by Georg **Schmitz** (ACCESS, Aachen) and V. Tikare (Sandia), who showed how the phase-field method can now simulate complex coupled fields leading to dendritic growth. E. Zipper (Katowice) introduced a simple model for cooperative phenomena in magnetic materials based on stacked rings that was of relevance to the work on manganites by

Bogdan **Dabrowski** (DeKalb), while John **Goodenough** (Austin) added his insights on transport in cooperativity.

For electronically active supermaterials, magnetic field studies (Qi Li, State College, and Françoise **Damay**, Imperial College) continue to dominate characterization activities, in combination with high pressure (P. Toulemonde, Grenoble), but the probing of electronic states by E. Kurmaev (Yekaterinburg) using x-ray spectroscopy is equally fundamental. Other photonic interactions were shown in the surprising phenomenon of persistent photoconductivity, another super property, in the research by A. Gilabert (Nice), and subpicosecond devices investigated by R. Sobolewski (Rochester). Davor **Pavuna** (Lausanne) explained how one engineers the band gap of materials to achieve such responses, as proved by the recent results on pulsed laser deposition reported by D. Blank (Twente).

Processing to tailor microstructure is a cross-cutting theme for "super"-materials. X. de la Fuente (Zaragoza) showed an exotic laser processing technique akin to zone melting, while Rudi Cloots (Liège) suggested several outside-the-box techniques to bring greater order to superconducting grains, notably a method inspired by closed-cell foams. At MIT, P. J. Ferreira studies magnetic field alignment of superconductor grains in molten matrix, and in a poster discussed the exciting field of magnetically activated shape-memory alloys for unintrusive in vivo actuators. In Japan, Jun **Takada** (Okoyama) is using electrochemistry, e.g. for Li-doping whence to enhance superconductivity. P. Diko (Kosice), G.D. Gu (Sydney), and Andrzej **Majchrowski** (Warsaw) showed thought-provoking results drawn from the crystal growth observations and technology, while D. Caplin (Imperial College) introduced us to the mysterious angular dependence of radiation treatments of superconductor tapes. Georges Schmitz made a surprise announcement of a breakthrough in HiTcS processing at ACCESS, namely the use of a fabric-like mask through which cuprate material can be microstructurally patterned to pin flux.

Finally, synthetic routes to obtain superproperties challenge chemists from both inorganic and organic sides of the discipline. Y.-W. Park (Seoul) has identified the essential junction between polyacetylene fibers that determines its properties as an organic conductor. Very certainly polyacetylene is a "super" material since the fact that it conducts at all is surprising. Phillipe **Odier** (Grenoble) introduced us to the biochemistry trick of gel-phase processing of superconductors as a way to control kinetics of reaction. This borrowing ideas across disciplinary boundaries is the essence of the materials research discipline. Rudi **Cloots** (Liège) pointed out the difficulty of synthesizing "super" compounds because of the negative interplay between charge carrier and phonon energies. This should be put in the same light as Janina **Molenda** (Krakow) report for Li intercalation in general transition xii

metal compounds.

In final remarks, V. Kresin (Lawrence Berkeley) pointed out the continuing lack of a truly comprehensive theory of high temperature superconductivity as the outstanding question (still) after 14 years of HiTcS existence. He noted with irony the re-emergence of mercury as an element of interest in the superconductor family, now numbering thousands (Hg was the first superconductor discovered). Maybe our views should be broadened by the discovery of the special behaviour of power law exponents for the electrical resistivity fluctuations in a magnetic field as discussed by Marcel **Ausloos** (Liège) and Marek **Pękała** (Warsaw). This might put some weight in favor of the *SO(5)* model of superconductivity.

Next for the "giant" magnetoresistive materials, we should note the hole doping effect on thermoelectric properties by Sawako **Nakamae** (Saclay), surely to be put in parallel to the same type of studies and measurements in superconducting cuprate ceramics by V. Guillen (Saclay). New "colossal" magnetoresistive material compounds were discussed by R.-S. **Liu** (Taipei).

Somewhat less fashionable emphasized Ivan **Nedkov** (Sofia) but of crucial importance for technology are the "simple" magnetic oxides. Zbigniew **Kletowski** (Wrocław) and Cristiana **Grigorescu** (Bucharest) surely agreed with that and the relevance of "super" properties for the ... "semi"-metallic and "semi"-magnetic conductors, hard or soft, from a magnetic point of view, but "super"interesting nevertheless.

One continues to ask, how is it that carriers or other agents in a material conspire to give surprising "super"properties? By broadening discussion to the category of supermaterials, as SUPERMAT (and SMART 99) have done, we have gained new paradigms to study this question.

Very warm and profound **acknowledgments** are in order here. SUPERMAT (and SMART 99) were made possible by NATO, the Regional Council of Provence-Alpes-Côte d'Azur, the Universities of Provence and Liège, and the individual institutions and sponsors of the participants. For Rudi Cloots and Marcel Ausloos, the Actions de Recherche Concertées (ARC 94-99/174) of the Communauté Française de Belgique provided some fund. Partial support from Belgian-Polish Scientific Exchange Agreement (1999 - 2000) for M. Pękała was helpful during preparation of SUPERMAT and proceedings. For Alan Hurd funding was provided in part by Sandia National Laboratories under contract to the Department of Energy, contract DE-AC04-94AL850000.

Rudi Cloots, Marcel Ausloos,
Marek Pękała, Alan J. Hurd, Gilbert Vacquier

ELECTRICAL PROPERTIES OF METAL-INSULATOR-SEMICONDUCTOR DEVICES WITH HIGH PERMITTIVITY GATE DIELECTRIC LAYERS

M. HOUSSA [(1,*)], P.W. MERTENS [(1)], M.M. HEYNS [(1)], AND A. STESMANS [(2)]

[(1)] *IMEC, Kapeldreef 75, B-3001 Leuven, Belgium.*
[(2)] *Department of Physics, Katholieke Universiteit Leuven, Celestijnelaan 200D, B-3001 Leuven, Belgium*

The electrical properties of metal-oxide-semiconductor capacitors with very thin high permittivity gate dielectric layers like Si_3N_4, Ta_2O_5, and Al_2O_3 are investigated. The field and temperature dependence of the gate current of these layers is analysed in order to determine the electrical conduction mechanisms in these layers, as well as to extract important physical parameters like the tunneling barrier height, electron effective mass, electron trap density and energy levels of the trapping centers. The time-dependent dielectric breakdown of these layers is next investigated. It is shown that soft breakdown occurs in these materials, like in ultra-thin SiO_2 layers. A percolation model is proposed to explain this phenomenon.

1. Introduction

The future generations of electronic devices will make full use of deep sub-micron CMOS (complementary metal oxide semiconductor) transistors. This schrinking of device dimensions is required to improve the performance and the packaging density of these systems. Currently, the most widely used insulating barrier in MOS field effect transistors is SiO_2. The reduction of CMOS device dimensions imply the increase of the gate capacitance, i.e. the reduction of the SiO_2 layer thickness. For instance, the operation of 100 nm (channel length) transistors will require a silicon dioxide layer with a thickness of less than 2 nm [1]. However, in such ultra-thin SiO_2 layers, the tunneling current becomes quite high, i.e. of the order of 0.1 to 1 A/cm^2 at 1.5 V. Such a high gate leakage current has a major impact on stand-by power consumption. Besides, as the gate oxide thickness is scaled down, the reliability of the insulating layer is much reduced [2,3]. It has been shown recently [3,4] that the temperature acceleration effect on the degradation of the gate oxide becomes quite important for SiO_2 layers thinner than 3 nm. Typically, the intrinsic time-to-breakdown of a 2 nm SiO_2 layer is reduced by 4 orders of magnitude from room temperature up to 150 °C. Consequently, the common criterion for oxide reliability, i.e. 10 years lifetime at operating voltage, should not be satisfied for SiO_2 layers thinner than 2.5 nm.

[(*)] Present address: Department of Physics, Katholieke Universiteit Leuven, Celestijnelaan 200 D, B-3001 Leuven, Belgium

R. Cloots et al. (eds.), Supermaterials, 1–20.
© *2000 Kluwer Academic Publishers. Printed in the Netherlands.*

An alternative way of increasing the gate capacitance is to use thicker insulating layers with higher dielectric constant than SiO_2. One would then expect to reduce the leakage current in the gate dielectric as well as to improve the gate insulator reliability when compared to SiO_2 layers with equivalent electrical thickness (as far as gate capacitance is concerned). Consequently, alternative gate dielectrics are currently widely investigated for future generation of MOS (metal oxide semiconductor) transistors [5-10]. Examples of such materials are Si_3N_4, Al_2O_3, ZrO_2, Ta_2O_5 and TiO_2.

In this work, we have investigated the electrical properties of MOS capacitors with thin high permittivity gate dielectric layers like Si_3N_4, Al_2O_3 and Ta_2O_5. High frequency capacitance-voltage measurements are performed in order to extract the dielectric constant of these materials, their equivalent SiO_2 thickness, the flat-band voltages, and density of electron traps. The field and temperature dependence of the gate current density is next studied in order to determine the electrical conduction mechanisms in these layers. The time-dependent dielectric breakdown of these materials is investigated using constant gate voltage stress of the capacitors. It is shown that soft breakdown is observed in these materials, like in ultra-thin SiO_2 layers. We suggest that this phenomenon is related to the formation of a percolation path between the electrons traps generated during the electrical stress of the gate dielectric layer.

2. Electrical characteristics of thin high permittivity gate dielectric layers

2.1 SILICON NITRIDE

Silicon nitride has a dielectric constant $\varepsilon=7.2$ and an energy band gap $E_G =5.1$ eV. This material is currently investigated [5,6] for the replacement of silicon dioxide in the (near) future generations of MOS devices, i.e. for the 100 nm channel length transistors. This material has the advantage of being compatible with the IC industry, and also prevents the penetration of B into the gate dielectric during p-type poly-Si deposition and activation.

Silicon nitride layers with thicknesses of 3 nm were deposited using a SiH_4/N_2 Remote Plasma Enhanced CVD (PECVD) process on p-type Si substrates [11]. The wafers were subsequently annealed in NO or N_2O at 900 °C during 30 s in order to reduce the bulk trap density in the Si_3N_4 layers. Capacitor structures with phosphorus-doped poly-Si layers were patterned on the wafers using wet lithography. The front side and backside of the wafers were Al-metalized in order to improve the contacts during the electrical measurements.

The nitrogen and oxygen content of these layers as measured from XPS (X-ray photo-emission spectroscopy) measurements [12] is given in Table 1. The N content in the layer is much decreased after the N_2O post anneal.

TABLE 1. N and O content in the Si_3N_4 layers after NO and N_2O post-anneals as determined from XPS measurements.

Sample	Post-anneal ambient	N content (at. %)	O content (at. %)
3 nm Si_3N_4	NO	29	35
3 nm Si_3N_4	N_2O	15	54

The high-frequency C-V characteristics of capacitors with Si_3N_4 gate dielectric layers are presented in Fig. 1 (a). One observes that the accumulation capacitance is higher for the NO post-annealed layer, as compared to the N_2O post-annealed layer. This result comes from the fact that the dielectric constant of Si_3N_4 decreases as the O content in the layer increases (cf. Table 1). From the accumulation capacitance, it is found that $\varepsilon=5.7$ for the NO post-annealed layer and $\varepsilon=4.8$ for the N_2O post-annealed layer respectively. The flat-band voltage of the capacitors is found to be -0.7 and -0.9 V for the NO post-annealed and N_2O post annealed layer. Since the workfunction difference between the p-type substrate and the n^+ doped poly-Si gate is expected to be around -0.7 eV, negative charges are present in the N_2O post-annealed layer. From the flat-band voltage shift, the density of these fixed negative charges is estimated to be 5×10^{18} cm^{-3}.

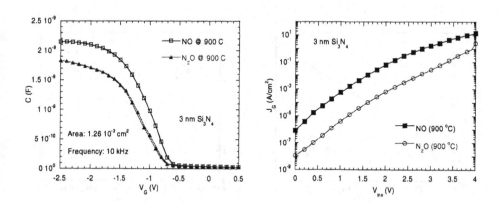

Figure 1. (a) Capacitance-voltage characteristics and (b) current-voltage characteristics of the 3 nm Si_3N_4 layers post-annealed in NO and N_2O.

4

The current density J_G - insulator voltage V_{ins} characteristics of the Si_3N_4 layers are shown in Fig. 1 (b). One observes that the current-density is lower in the N_2O post-annealed layer, i.e. in the layer containing the highest concentration of O. This result can be explained by the fact that the barrier height ϕ for electron tunneling from the gate to the substrate is increased as the O content in the Si_3N_4 layer is increased, ϕ lying between 2.1 eV in pure silicon nitride [13] and 3.2 eV in pure silicon dioxide [13]. The barrier height in the NO and N_2O post-annealed layer will be extracted below.

The current density of the silicon nitride layers has been analysed using a tunneling model based on the WKB approximation. The expression for the direct tunneling (DT) current is given by [14]

$$ J_{DT} = \left(\frac{64 \, m_t \, e}{\pi^2 h^5 c_1^2} \right) \left(\frac{t_{ins}}{c_2} \right)^2 \left(\frac{m^*}{\sqrt{m_z}} \right)^2 (c_3 - c_4) \exp(-c_2) \tag{1} $$

where m_t and m_z are the transverse and normal electron effective mass in the poly-Si gate, m^* the electron effective mass in the gate insulator, t_{ins} the gate insulator thickness and c_1, c_2, c_3 and c_4 are functions of the insulator voltage V_{ins}, electron effective mass m^*, insulator band gap E_G and tunneling barrier height ϕ [14].

The fit to the current density of the NO post-annealed layer using Eq.(1) is shown in Fig. 2 (a). The free parameters are the electron tunneling barrier height and the electron effective mass m^*. One can see that the data are quite well reproduced by the DT model for $V_{ins} < \phi$, and the electron effective mass and barrier height are found to be $m^* = 0.48 \, m_0$ and $\phi = 2.3$ eV respectively.

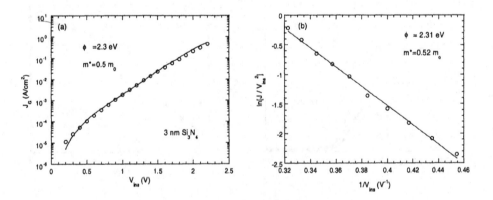

Fig.2. Current-voltage characteristics of the 3 nm Si_3N_4 layer post-annealed in NO and theoretical curves obtained from Eq.(1) and (2) in (a) direct tunneling and (b) Fowler-Nordheim tunneling regime respectively.

The current density – gate insulator voltage characteristics of the NO post-annealed Si_3N_4 layer are presented on a Fowler-Nordheim (FN) plot in Fig. 2 (b) for $V_{ins}>2.3$ V. The solid line in Fig. 2 (b) is a fit to the data using the FN tunneling current [15]

$$J_{FN} = A \left(\frac{V_{ins}}{t_{ins}} \right)^2 \exp \left(\frac{-Bt_{ins}}{V_{ins}} \right) \qquad (2)$$

where A and B parameters are related to the effective mass in the insulator m* and the barrier height ϕ [15]. As shown in Fig. 2 (b), the experimental results follow approximately a straight line as predicted by Eq. (2). The values of the effective mass and barrier height are respectively 0.52 m_0 and 2.31 eV, values consistent with those found from the fit in the DT regime.

The tunneling barrier height as a function of O concentration in the Si_3N_4 layer is presented in Fig. 3. One observes that ϕ increases from 2.1 eV (pure silicon nitride) to 3.2 eV (pure silicon dioxide) with increasing the oxygen content in the Si_3N_4 layer. Consequently, oxide-rich Si_3N_4 layers (or nitrided silicon dioxide) will have a lower gate leakage current than nitride-rich Si_3N_4 layers, provided that layers with the same physical thickness are compared. This results explains the lower gate current observed in the N_2O post-annealed layer, see Fig. 1 (b).

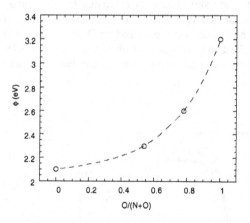

Fig.3. Electron tunneling barrier height of Si_3N_4 as a function of the oxygen content in the layer. These results were obtained by fitting the gate current density of different layers in the direct and Fowler-Nordheim tunneling regimes.

6

2.2. TANTALUM PENTOXIDE

This material has a dielectric constant of the order of 25 and an energy band gap of 4.4 eV. Ta_2O_5 has been investigated as an alternative dielectric for both DRAM (dynamic random access memory) and MOS field effect transistors [7-9, 16]. The high dielectric constant of this material makes it a promising candidate as an alternative gate insulator. However, this material is not stable on Si, i.e. an interfacial SiO_2 layer is grown during the deposition of Ta_2O_5 on Si substrates. This interfacial layer reduces the total gate capacitance, and one has to find a way to control/reduce its growth during the deposition of the high permittivity dielectric layer.

N-type Si wafers were first pre-annealed in NO at 800 °C during 20 sec. During this anneal, an oxynitride layer of the order of 1 nm is grown on the Si substrate. This ultra-thin interfacial layer improves the quality of the Si/dielectric interface, i.e. reduces the interface trap density. Ta_2O_5 layers with thicknesses of 6 nm, 8 nm and 10 nm were next deposited by MOCVD (metal-organic chemical vapor deposition) at 450 °C. Part of the wafers were post-annealed in O_2 at 600 °C during 30 sec in order to reduce the trap density and carbon content in the Ta_2O_5 layers. A low temperature anneal was chosen in order to avoid the crystallization of the Ta_2O_5 layers, which occurs at around 700 °C [16]. MOS capacitor structures with Pt top electrodes were patterned using lithography and Pt dry etching.

The high frequency capacitance-voltage characteristics of $SiON/Ta_2O_5$ gate stacks with 6 nm Ta_2O_5 layers are shown in Fig. 4 (a). One observes that the capacitance is lower in the post-annealed layer as compared to the as-deposited stack, which is due to the increase of the interfacial oxynitride layer thickness during the rapid thermal anneal in O_2 (see below). The flat-band voltage is estimated to be around +0.6 V for the as-deposited and post-annealed layers respectively, which is close to the workfunction difference between the n-type Si substrate and the Pt electrode, i.e. 0.6-0.7 eV. It should be noticed that the flat-band voltage shift observed during double voltage sweep, i.e. from accumulation to inversion, and back to accumulation, is of the order of 40 mV.

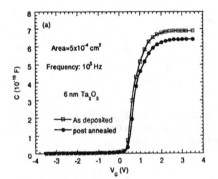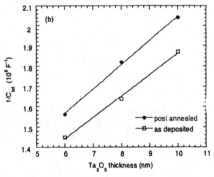

Fig. 4. (a) High frequency C-V characteristics of $SiON/Ta_2O_5$ gate stack with a 6 nm Ta_2O_5 layer. (b) Inverse of the total gate capacitance of $SiON/Ta_2O_5$ gate stack as a function of Ta_2O_5 layer thickness. Solid lines are fit to the data using Eq.(3).

This flat-band voltage shift is due to the contribution of charges trapped in the SiON/Ta$_2$O$_5$ gate stack during the first voltage sweep; the density of these trapped charges is estimated to be about 4×10^{17} cm^{-3}.

Considering a simple series capacitance model, the total capacitance C$_{tot}$ of the SiON/Ta$_2$O$_5$ gate stack is given by

$$\frac{1}{C_{tot}} = \frac{t_{Ta_2O_5}}{A\varepsilon_o\varepsilon_{Ta_2O_5}} + \frac{t_{SiON}}{A\varepsilon_o\varepsilon_{SiON}} \tag{3}$$

where t_{Ta2O5} and t_{SiON} are the thicknesses of the Ta$_2$O$_5$ and SiON layer respectively, ε_{Ta2O5} and ε_{SiON} the dielectric constants of Ta$_2$O$_5$ and SiON respectively, ε_o the permittivity of free space and A the capacitor area. According to Eq.(3), a plot of $1/C_{tot}$ as a function of the Ta$_2$O$_5$ layer thickness should give a straight line from which ε_{Ta2O5} and t_{SiON} can be extracted. The inverse of the total gate capacitance of the as-deposited and post-annealed gate stacks is presented in Fig. 4 (b) as a function of tantalum pentoxide thickness. One can see that the data follows approximately a straight line on such a plot, as predicted by Eq.(3). Assuming that the dielectric constant of the oxynitride layer is 3.8, one finds ε_{Ta2O5}=21.4 and t_{SiON}=1.2 nm for the as-deposited layers, and ε_{Ta2O5}=20.5 and t_{SiON}=1.4 nm for the post-annealed layers. The thickness of the interfacial layer is thus slightly increased after the rapid thermal anneal in O$_2$, which is consistent with the lower accumulation capacitance value observed in the post-annealed gate stack, cf. Fig. 4 (a).

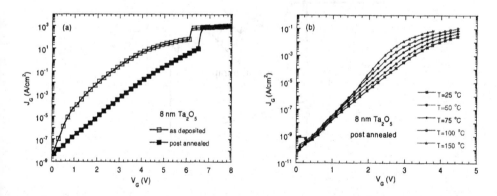

Fig.5. (a) Current density J$_G$ as a function of gate voltage V$_G$ of the as-deposited and post-annealed SiON/Ta$_2$O$_5$ gate stack with 8 nm Ta$_2$O$_5$ layers. (b) J$_G$ vs. V$_G$ of the post-annealed layer from room temperature to 150 °C.

The current density J$_G$ vs. gate voltage V$_G$ of capacitors with 8 nm Ta$_2$O$_5$ layers is shown in Fig. 5 (a). One can see that J$_G$ is reduced by 3 orders of magnitude after the post-anneal in O$_2$. This indicates that the rapid thermal anneal treatment is efficient in

8

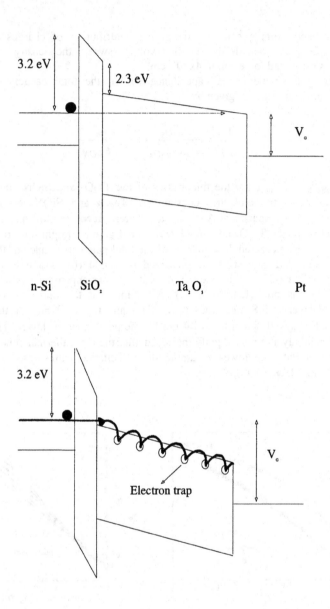

Fig. 6. Schematic energy band diagram of the $Si/SiO_2/Ta_2O_5/Pt$ system at low applied voltage (top figure) and high applied voltage (bottom figure).

reducing the density of bulk traps in the Ta_2O_5 layer. The increase of the total gate stack thickness resulting from the post-anneal in O_2 is also responsible for the decrease of the gate current density.

The temperature dependence of the current-voltage characteristics of the post-annealed gate stack with 8 nm Ta_2O_5 layer is presented in Fig. 5 (b), from 25 °C to 150 °C. One observes that J_G is weakly temperature dependent in the low gate voltage region, i.e.

below about 1.5 V, while the temperature dependence of J_G becomes more important at larger gate bias. This suggests that the electrical conduction is dominated by tunneling through the gate stack at low voltages, since tunneling is slightly temperature dependent [17], and that Poole-Frenkel conduction in the Ta_2O_5 layer becomes important at larger gate bias. The conduction through the $Si/SiO_2/Ta_2O_5/Pt$ system at low and high applied gate voltage is illustrated in Fig.6 (a) and (b) respectively.

The Poole-Frenkel conduction mechanism results from the field-enhanced thermal excitation of trapped electrons in the conduction band of the gate insulator [18]. For traps with a coulombic potential well, the expression of the Poole-Frenkel current is given by

$$J = C E \exp\left(\frac{-q\phi_B}{k_B T}\right) \exp\left(\frac{q\sqrt{qE/\pi\varepsilon_0 n^2}}{k_B T}\right) \tag{4}$$

where ϕ_B is the depth of the trap potential well, C is a constant depending on the trap density and n is the refractive index of the material.

In order to determine the field and temperature dependence of the current in the Ta_2O_5 layers, the electric field in Ta_2O_5 and in the interfacial SiON layer have to be calculated. Assuming that no charge is accumulated at the Ta_2O_5/SiON interface, the electric field across the 2 layers is respectively given by [18],

$$E_{SiON} = \frac{V_{ins}}{\frac{\varepsilon_{SiON}}{\varepsilon_{Ta_2O_5}} t_{Ta_2O_5} + t_{SiON}} \tag{5}$$

$$E_{Ta_2O_5} = \frac{V_{ins}}{\frac{\varepsilon_{Ta_2O_5}}{\varepsilon_{SiON}} t_{SiON} + t_{Ta_2O_5}} \tag{6}$$

where $V_{ins} = V_G - V_{FB}$ is the voltage applied to the gate dielectric stack, ε_{SiON} and $\varepsilon_{Ta_2O_5}$ are the dielectric constant of the interfacial layer and Ta_2O_5 layer respectively, t_{SiON} and $t_{Ta_2O_5}$ the thickness of the interfacial layer and Ta_2O_5 layer respectively.

The current-voltage characteristics of the $SiON/Ta_2O_5$ gate stacks with 6 nm Ta_2O_5 layers are shown in Fig. 7 (a) on a so-called Poole-Frenkel plot, i.e. $J_G/E_{Ta_2O_5}$ is presented as a function of $E_{Ta_2O_5}^{1/2}$ on a semi-log plot. The data is presented for $V_G > 2$ V, i.e. in the voltage region where the temperature dependence of the gate current becomes stronger (cf. Fig. 5). The field in the Ta_2O_5 layer has been calculated from Eq.(6), with t_{SiON} and ε_{SiON} as determined from the analysis of the high frequency C-V characteristics, cf. Fig. 4 (b). One observes that the data for the as-deposited and post-annealed layers can be quite well fitted by Eq.(4), indicating that the conduction

10

mechanism in the Ta₂O₅ layer is most probably of Poole-Frenkel type. The values of the refractive index n obtained from the fits to the data are n=2.26 for the as-deposited layer and n=2.44 for the post-annealed layer respectively. These values are in good agreement with those reported in the literature [16].

The gate current density of the gate stacks with 6 nm Ta₂O₅ layers at a fixed gate voltage of 3V (corresponding to an electric field of 2.2 MV/cm in the tantalum pentoxide layer) is presented in Fig. 7 (b) as a function of 1/T on a semi-log plot (from 75 °C to 150 °C). One can see that the data follows approximately a straight line in such a plot, as predicted by Eq.(4). Consequently, the temperature dependence of the current in the Ta₂O₅ layer above 3V is consistent with the Poole-Frenkel mechanism. From the slopes of the straight line fits shown in Fig. 7 (b), the trap energy level ϕ_B is found to be 0.84 eV for the as-deposited layer and 0.85 eV for the post-annealed layer.

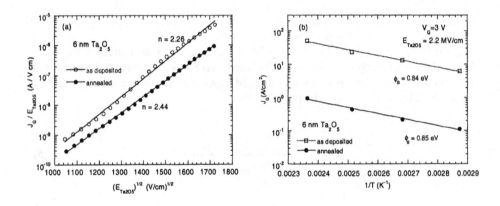

Fig.7. (a) Poole-Frenkel plot of the current-voltage characteristics of the gate stack with 6 nm Ta₂O₅ layers. Solid lines are fits top the data using Eq. (4). (b) Temperature dependence of the gate current density of the gate stack with 6 nm Ta₂O₅ layer at V_G=3 V. Solid lines are fits to the data using Eq. (4).

2.3 ALUMINIUM OXIDE

Aluminium oxide has a dielectric constant of the order of 10 and a band gap of the order of 8 eV [13]. Though its dielectric constant is not very high, this material is investigated as a potential alternative gate dielectric [19] due to its large band gap and high barrier height on Si, which is of the order of 3 eV. This value is close to the barrier height of Si/SiO_2 (3.2 eV).

Al_2O_3 layers with thicknesses of 7.5 nm and 15 nm were deposited by ALCVD (atomic layer chemical vapor deposition) at 300 °C on n-type Si wafers. The wafers received a RCA clean prior to the gate dielectric deposition, i.e. a native SiO_2 interfacial layer with thickness of about 1.5 nm was present before Al_2O_3 depostion. Gold dots ($5x10^{-3}$ cm^2 area) were evaporated on the wafers through a shadowmask in order to form MOS capacitor structures.

The high-frequency C-V characteristics of the SiO_2/Al_2O_3 gate stack with 7.5 nm Al_2O_3 layer recorder during double voltage sweep are shown in Fig. 8 (a). A flat-band voltage shift of the order of 0.35 V is observed between the 2 sweeps, which results from the charges trapped in the SiO_2/Al_2O_3 stack after the first voltage sweep. The density of these trapped charges is of the order of $1.4x10^{18}$ cm^{-3}.

In order to determine the thickness of the SiO_2 interfacial layer as well as the dielectric constant of the Al_2O_3 layer, the inverse of the total gate capacitance is plotted in Fig. 8 (b) as a function of the aluminium oxide layer thickness, cf. Eq.(3). The interfacial SiO_2 layer thickness is estimated to be 1.9 nm and the dielectric constant of the Al_2O_3 layer is found to be 7.9.

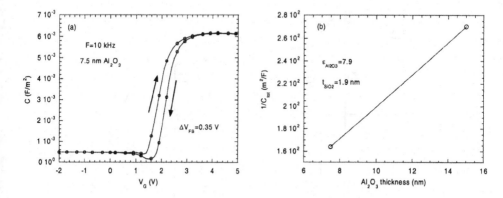

Fig. 8. (a) High frequency C-V characteristics of SiO_2/Al_2O_3 gate stacks with 7.5 nm Al_2O_3 layer recorded during double voltage sweep, i.e. from accumulation to inversion and back to accumulation. (b) Inverse of the total gate capacitance vs. Al_2O_3 layer thickness. The solid line is a fit to the data using Eq. (3).

12

The gate current density of the gate stacks with 7.5 nm and 15 nm Al$_2$O$_3$ layers is shown in Fig. 9. as a function of the voltage across the gate stack V$_{ins}$. The behaviour of the gate current is quite similar to the one observed in thin SiO$_2$ layers [15], i.e. the conduction mechanism is likely tunneling from the Si substrate through the gate stack, eventually through the traps present in the Al$_2$O$_3$ layer (so called trap-assisted tunneling mechanism).

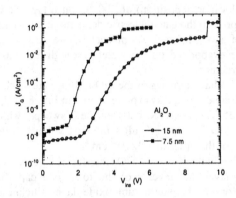

Fig. 9. Current-voltage characteristics of the SiO$_2$/Al$_2$O$_3$ gate stack with 7.5 nm and 15 nm Al$_2$O$_3$ layers.

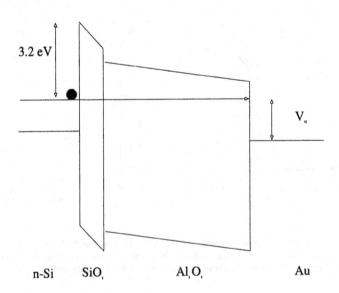

Fig. 10. Schematic energy band diagram of the n-Si/SiO$_2$/Al$_2$O$_3$/Au system.

The schematic energy band diagram of the Si/SiO$_2$/Al$_2$O$_3$/Au system is shown in Fig. 10. The Si/Al$_2$O$_3$ barrier height is of the order of 3 eV [13], so that the energy band discontinuity between SiO$_2$ and Al$_2$O$_3$ is quite small, i.e. about 0.2 eV. Consequently, the most probable conduction mechanism is tunneling through the SiO$_2$ and Al$_2$O$_3$ stack even at high voltage, unlike the situation in the SiO$_2$/Ta$_2$O$_5$ layer were the large band discontinuity between both gate insulators (of the order of 2.3 eV, see Fig. 6) leads to direct tunneling in the SiO$_2$ layer and Poole-Frenkel conduction through the Ta$_2$O$_5$ layer when V$_{ox}$ is greater than q$\phi_{Si/Ta2O5}$=0.9 eV.

The tunneling current through the SiO$_2$/Al$_2$O$_3$ stack has been calculated within the WKB approximation. The current density is given by [20]

$$J(V) = \frac{q\, m^*(k_B T)}{2\pi^2 \hbar^3} \int_{o}^{E_F} T(E,V) \ln\left(\frac{1+\exp[(E_F - E)/k_B T]}{1+\exp[(E_F - E - qV)/k_B T]} \right) dE \qquad (7)$$

where $T(E,V)$ is the transmission probability, E the electron energy, E_F the Fermi level at the cathode interface, and V the voltage applied across the gate stack. The transmission probability is the product of the transmission probability in the SiO$_2$ layer and the Al$_2$O$_3$ layer, i.e.

$$T(E,V) = T_{SiO2}(E,V_{SiO2}) * T_{Al2O3}(E,V_{Al2O3}) \qquad (8)$$

where V$_{siO2}$ and V$_{Al2O3}$ are calculated according to Eq.(5) and (6). The transmission probabilities in SiO$_2$ and Al$_2$O$_3$ are calcultated within the WKB approximation, both in the direct and Fowler-Nordheim tunneling regimes (see for example ref. [15]). The free parameters are the barrier height of Si/SiO$_2$ and Si/Al$_2$O$_3$ as well as the electron effective masses in both dielectrics.

The current density of the SiO$_2$/Al$_2$O$_3$ gate dielectric stack with 7.5 nm Al$_2$O$_3$ layer is shown in Fig. 11. The solid line is a fit to the data obtained by using Eq.(7) and (8). A good agreement between the data and theoretical curve is obtained with the following realistic values of the free parameters: $\phi_{Si/SiO2}$=3.2 eV, $\phi_{Si/Al2O3}$=2.9 eV, m*$_{siO2}$=0.5 m$_o$ and m*$_{Al2O3}$=0.3 m$_o$, where m$_o$ is the free electron mass.

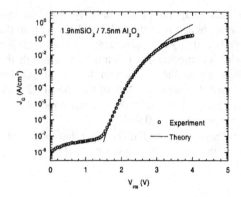

Fig. 11. Current-density vs gate insulator voltage of the SiO_2/Al_2O_3 gate stack with 1.9 nm SiO_2 layer and 7.5 nm Al_2O_3 layer. The solid line is a fit to the data obtained by using Eq.(7) and (8).

3. Soft breakdown in very thin high permittivity gate dielectric layers

Dielectric breakdown of thin gate dielectric layers is considered as a major reliability issue in integrated circuits [21-23]. It is well known that in SiO_2 layers thinner than 5 nm, soft breakdown is observed during constant voltage or constant current stress of MOS devices [24-26]. Soft breakdown corresponds to a large increase of the gate current in the low gate voltage range as well as to the occurrence of fluctuations in the time-dependence of the gate voltage or gate current. Soft breakdown is now recognized as a major device failure mode in MOS field effect transistors.

The time-dependence of the gate current I_G of $SiON/Ta_2O_5$ gate stacks with 6 nm and 8 nm Ta_2O_5 layers observed during constant voltage stress of MOS capacitors is shown in Fig. 12 (a) [27]. One observes that the current rapidly increases, due to the generation of positive charges in the $SiON/Ta_2O_5$ gate stack. After 202 s (8 nm Ta_2O_5) and 142 s (6 nm Ta_2O_5), the current suddenly increases (jump of the order of 5 to 10 %), followed by the occurrence of fluctuations. A much larger jump in the gate current is observed after 294 s and 372 s for the 6 nm and 8 nm layer respectively, corresponding to the complete breakdown of the gate dielectric. In Fig. 12 (b), the current-voltage characteristics of the $SiON/Ta_2O_5$ gate stack with 6 nm Ta_2O_5 layer are shown after different time during constant voltage stress of the capacitors. A sudden large increase of the gate current in the low gate voltage region is observed at the time corresponding to the occurrence of fluctuations in the time-dependent current, see Fig. 12 (a). These fluctuations and increased gate current are characteristic features of the soft breakdown phenomena reported for ultra-thin SiO_2 layers [24-26].

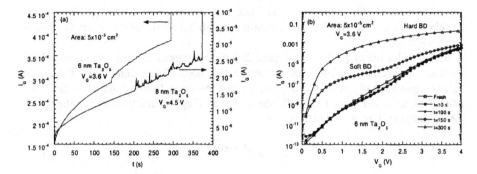

Fig. 12. (a) Time-dependence of the gate current of SiON/Ta$_2$O$_5$ gate stacks during constant voltage stress of MOS capacitors. (b) Current-voltage characteristics of the SiON/Ta$_2$O$_5$ stack with 6 nm Ta$_2$O$_5$ layer recorded after different time of the constant voltage stress.

It has been shown recently that the gate current behaves like a power law of the applied gate voltage after soft breakdown of ultra-thin SiO$_2$ layers [25,28]

$$V_G \propto I_G^\alpha \tag{9}$$

with $\alpha = 0.32 \pm 0.02$. This power law behaviour of the gate current after soft breakdown can be explained by the percolation theory of non-linear conductor networks with a distribution of percolation thresholds [29]. During the stressing of MOS capacitors, electrons traps are created in the gate insulator layer. These traps randomly occupy the sites of the insulator "lattice" as shown schematically in Fig.13. We assume that the current between two neighbour traps is proportional to the square of the applied voltage, i.e.

$$i = \sigma_{ab} v^2 \tag{10}$$

where σ_{ab} is the bound conductivity. This quadratic behaviour comes from the product of the trapped charge density in the capacitor $\rho_t \, \alpha \, v$ and the electron drift mobility $\mu_e \, \alpha \, v$ [30]. When a critical number of traps is generated in the gate insulator, a percolation path is formed between the gate electrode and the Si substrate [31], leading to a sudden increase of the current in the system. This situation corresponds to the occurrence of soft breakdown in the gate oxide. Since gate dielectric layers are very thin, one expect that the percolation threshold (i.e. the critical number of traps needed to trigger soft breakdown) is not a single value, but rather corresponds to a distribution of percolation thresholds induced by the finite size of the system [29,32]. This is also correlated to the wide distribution of time-to-quasi-breakdown values observed in ultra-thin SiO_2 layers [31].

Roux and Herrmann [29] studied percolation phenomena in finite size disordered systems with such a distribution of percolation thresholds. They found that if the current behaves quadratically with the applied voltage between two individual sites of the lattice (as assumed in our model), the current-voltage characteristics of the whole (percolating) system should behave like

$$V \propto I^{0.37 \pm 0.04} \qquad (11).$$

This power-law dependence of the current is consistent with the experimental results observed in ultra-thin gate SiO_2 layers [32].

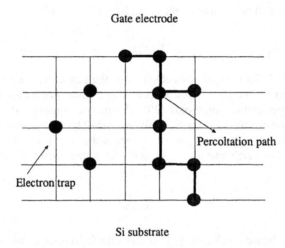

Fig. 13. Schematic illustration of the percolation path formed between the electron traps generated in the gate dielectric layer during electrical stress of the capacitor.

The current-voltage characteristics of SiON/Ta$_2$O$_5$ gate stacks after the occurrence of soft breakdown are presented in Fig. 14 (a) on a log-log scale. The data observed on a SiO$_2$/Al$_2$O$_3$ stack with 7.5 nm Al$_2$O$_3$ layer after soft breakdown are presented in Fig. 14 (b). One observes that the gate current behaves like a power law, as predicted by Eq.(11). The exponent α is found to lie between 0.2 and 0.36, in reasonable agreement with the values predicted by the percolation model, see Eq.(11).

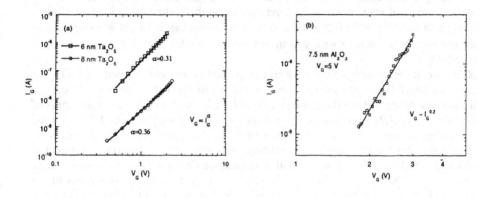

Fig. 14. Gate current vs. gate voltage of (a) SiON/Ta$_2$O$_5$ layers and (b) SiO$_2$/Al$_2$O$_3$ layers on a log-log scale.

4. Conclusions

The electrical properties of metal-insulator-semiconductor capacitors with high permittivity gate dielectric stacks have been investigated. First, the effect of different post-annealing treatment on the electrical characteristics of ultra-thin remote-plasma CVD Si_3N_4 layers has been studied. It has been shown that the accumulation capacitance decreases as the oxygen content in the silicon nitride layer is increased, which results from the decrease of the effective dielectric constant ε of the material with increasing O concentration. On the other hand, the gate current density is lower (by about 2 orders of magnitude) in the layer containing the highest oxygen concentration. This result arises from the fact that the electron tunneling barrier height ϕ is increased as the oxygen content is increased, ϕ lying between 2.1 eV for pure Si_3N_4 and 3.2 eV for pure SiO_2.

Next, we have investigated the electrical properties of as-deposited and O_2-post annealed $SiON/Ta_2O_5$ gate stacks. From the analysis of the high-frequency C-V curves, it has been shown that the dielectric constant of Ta_2O_5 is about 20 and the thickness of the interfacial SiON layer is about 1.2 nm for the as-deposited stack and 1.4 nm for the post-annealed stack. The gate current density is decreased by about 3 orders of magnitude after the post-annealing treatment, indicating that this thermal treatment is efficient in reducing the density of bulk and interface electron traps in the Ta_2O_5 layer and at the $SiON/Ta_2O_5$ interface respectively. From the analysis of the current in the gate stack, it has been proposed that tunneling through both insultaing layers occurs at low voltage (below about 1.5 to 2 V), and that tunneling in the SiON layer and Poole-Frenkel conduction in the Ta_2O_5 layer takes place at higher voltage. The energy level of the traps in Ta_2O_5 and the refractive index of this material have been found to be 0.85 eV and 2.2-2.4 respectively.

The analysis of the high-frequency C-V characteristics of SiO_2/Al_2O_3 stacks deposited by atomic layer CVD has shown that the thickness of the interfacial layer is about 1.9 nm and the dielectric constant of Al_2O_3 is about 8. The current-voltage characterisitcs of SiO_2/Al_2O_3 has been modeled by considering the tunneling conduction through the gate stack. Reasonable agreement between the experimental results and the theoretical curve has been obtained, with a SiO_2/Al_2O_3 barrier height of 2.9 eV and an effective mass in the Al_2O_3 layer of 0.3 m_0.

Finally, the time-dependent dielectric breakdown of SiO_2/high permittivity dielectric gate stack has been studied, by stressing capacitors with constant voltage stress. It has been shown that fluctuations are observed in the gate current after a certain stress time, and that the gate current suddenly increases in the low gate voltage region after the occurrence of these fluctuations. These features are characteristics of the so-called soft breakdown phenomena. Besides, the gate current of the gate stack behaves like a power-law with the applied gate voltage after soft breakdown, like in ultra-thin SiO_2 layers. This power law behaviour can be explained by the percolation theory of (finite-size) systems with a distribution of percolation thresholds, assuming that a percolation path is formed between the traps generated in the gate insulator layer during the electrical stress of the capacitors.

References

1. The 1997 Semiconductor Industry Association Roadmap (Austin, Texas, 1997).

2. Stathis, J.H. and DiMaria, D.J. (998) *Techn. Dig. International Electron Devices Meeting*, (IEEE, Piscataway), 167.

3. Degraeve, R., Pangon, N., Kaczer, B., Nigam, T., Groeseneken, G., and Naem, A. (1999) *Technical Digest of VLSI Symposium* (IEEE, Piscataway), 59.

4. DiMaria, D.J. and Stathis, J.H. (1999) *Appl. Phys. Lett.* **74**, 1752.

5. Lucovsky, G., Niimi, H., Wu, Y., Parker, R.C., and Hauser, J.R. (1998) *J. Vac. Sc. Technol.* **A 16**, 1721.

6. Ma, T.P. (1998) *IEEE Trans. Electron Dev.* **45**, 680.

7. Cava, R.J. and Krajewski, J.J. (1998) *J. Appl. Phys.* **83**, 1613.

8. Houssa, M., Degraeve, R., Mertens, P.W., Heyns, M.M., Jeon, J.S., Halliyal, A., and Ogle, B. (1999) *J. Appl. Phys*, in press.

9. McKee, R.A., Walker, F.J., and Chisholm, M.F. (1998) *Phys. Rev. Lett.* **81**, 3014.

10. Mikhaelashvili, V., Betzer, Y., Prudnikov, I., Orenstein, M., Ritter, D., and Eisenstein, G. (1998) *J. Appl. Phys.* **84**, 6747.

11. Pomarede, C., Werkhoven, C., Weidmann, J., Bergman, T., Gschwandtner, A., and Houssa, M. (1999) *Proceedings of the MRS Spring Meeting*, in press.

12. Conard, T., De Witte, H., Vandervorst, W., Houssa, M., Heyns, M., Pomarede, C., and Werkhoven, C. (1999) to appear in the *Proceedings of the MRS Fall Meeting*.

13. Balk, P. (1988) in *The Si-SiO₂ system*, Ed. by P. Balk, Elsevier, Amsterdam, p.1.

14. Houssa, M., Degraeve, R., Pomarede, C., van Dijk, K., Werkhoven, C., Mertens, P.W., and Heyns, M.M. (1999) submitted to *J. Appl. Phys.*

15. Depas, M., Vermeire, B., Mertens, P.W., Van Meirhaeghe, R.L., and Heyns, M.M. (1995) *Solid-State Electron.* **38**, 1465.

16. Chaneliere, C., Autran, J.L., Devine, R.A.B., and Balland, B. (1998) *Mater. Sc. Eng. Rep.* **R22**, 269.

17. Pananakakis, G., Ghibaudo, G., Kies, R., and Papadas, C. (1995) *J. Appl. Phys.* **78**, 2635.

18. Sze, S.M. (1969) *Physics of Semiconductor Devices*, Wiley, New York.

19. Manchanda, L. *et al.* (1998) *Techn. Dig. International Electron Devices Meeting*, (IEEE, Piscataway), 605.

20. Fromhold, A.T. (1981) *Quantum Mechanics for Applied Physics and Engineering*, Dover, New York.

21. Lee, J.C., Chen, I.C., Hu, C. (1988) *IEEE Trans. Electron Dev.* **35**, 2268.

22. DiMaria, D.J., Cartier, E., Arnold, D. (1993) *J. Appl. Phys.* **73**, 3367.

23. Depas, M., Vermeire, B., and Heyns, M.M. (1996) *J. Appl. Phys.* **80**, 382.

24. Depas, M., Nigam, T., and Heyns, M.M. (1996) *IEEE Trans. Electron Dev.* **43**, 1499.

25. Okada, K. and Tanuguchi, K. (1997) *Appl. Phys. Lett.* **70**, 351.

26. Houssa, M., Vandewalle, N., Nigam, T., Ausloos, M., Mertens, P.W., and Heyns, M.M. (1998) *Techn. Dig. International Electron Devices Meeting*, (IEEE, Piscataway), 909.

27. Houssa, M., Mertens, P.W., Heyns, M.M., Jeon, J.S., Halliyal, A., and Ogle, B. (1999) *Solid-State Electron.*, in press.

28. Houssa, M., Nigam, T., Mertens, P.W., and Heyns, M.M. (1998) *Appl. Phys. Lett.* **73**, 514.

29. Roux, S. and Herrmann, H.J. (1987) *Europhys. Lett.* **4**, 1227.

30. Lampert, M.A. and Mark, P. (1970) *Current Injection in Solids*, Academic Press, New York.

31. Degraeve, R., Groeseneken, G., Bellens, R., Ogier, J.L., Depas, M., and Maes, H. (1998) *IEEE Trans. Electron Dev.* **45**, 904.

32. Houssa, M., Nigam, T., Mertens, P.W., and Heyns, M.M. (1998) *J. Appl. Phys.* **84**, 4351.

PHASE-FIELD MODELLING OF EVOLVING MICROSTRUCTURES

G. J. SCHMITZ
ACCESS e.V
Intzestr. 5, D-52072 Aachen, Germany

Abstract

Recent developments of the phase-field concept and its applications in modeling microstructures evolving during solidification of multicomponent and multiphase alloys are reviewed and future directions of the method like e.g. coupling to thermodynamic databases or coupling between macroscopic process simulation and simulation of microstructure evolution are highlighted.

1. Introduction

The phase field approach of modeling phenomena occurring during solidification and in solid/solid phase changes in the recent years has gained more and more importance. Being based on fundamental aspects of phase-transitions in general, research in the „phase field" by now has essentially been performed by mathematicians and theoretical physicists. Recent developments - e.g. the extension of the phase field concept to the description of multiphase equilibria and multiple alloyed systems - make this method a powerful tool for materials scientists, foundry engineers and any other people trying to optimize the quality of their products by optimizing the microstructure of the materials.
Scopes of this paper are to review the developments of the phase field concept, to define its actual state and to identify future directions for the development of the phase field concept to meet the tasks of materials scientists/engineers.

2. Fundamentals of the phase-field model

In the field of materials sciences especially the phase-field theory describing first order phase-transitions like solidification/crystallization has made significant progress. As the phase-field method is based on Time-Dependant Ginzburg Landau theories it provides a double well potential of the order parameter, which in solidification problems often is identified with the fraction ϕ of the solid phase varying between 0 and 1. In the early days of phase field modeling the two stable states were identified by order parameter values of +/-1. The general structure of the free energy formulation, from which the equations of motion are derived by functional differentiation, in the phase-field concept consist of three contributions :

R. Cloots et al. (eds.), Supermaterials, 21–26.
© 2000 *Kluwer Academic Publishers. Printed in the Netherlands.*

- a gradient energy being related to gradients of the order parameter
- a potential energy (in general a double well potential with minima either at +/-1 or - in more recent work - at 0 and 1. Sometimes a double obstacle potential is used.)
- a driving force term describing the deviation from equilibrium being proportional to an undercooling ΔT in case of solidification of pure substances or - in the more general case of alloy solidification - to the difference in Gibbs energies of the bulk phases.

Different models taking into account specific aspects (e.g. anisotropy) are based on specific selections/formulations/assumptions for each of these contributions. The stationary solution ($d\phi/dt =0$) of the simplest phase-field equation

$$\tau \, d\phi/dt = \varepsilon^2\nabla^2\phi + W'(\phi) + m(\Delta T) \tag{1}$$

in thermodynamic equilibrium ($m(\Delta T)= 0$) yields a hyperbolic tangent profile for the order parameter in case of the double well potential or a sine-like transition profile in case of $W(\phi)$ being a double obstacle potential. Both solutions are centered at position $\underline{x}$ $=0$ in space. In case of a non-vanishing driving force ($m(\Delta T)\neq 0$) these solutions move in space with a velocity proportional to the degree of undercooling preserving the profile shape. This motion in solidification problems is associated with a release of the latent heat of fusion appearing as a source term ($\sim L d\phi/dt$) in the heat conduction equation:

$$dT/dt = \lambda\nabla^2 T + L d\phi/dt \tag{2}$$

The non-linear coupling between the phase-field equation (1) and the heat-diffusion equation (2) via the contributions $L d\phi/dt$ and $m(\Delta T)$ is the basis for the evolution of complex structures like dendrites when numerically iterating these equations in time [1-9].
In case of alloy solidification additional couplings arise to the diffusion equations of the individual components, to stress/strain fields and to the Navier-Stokes equation of fluid flow[10-12]. Moreover, these equations have mutual couplings describing e.g. cross-diffusion , Soret and Dufor effects, etc.

During solidification of an alloy in many cases several phases are coexisting like e.g. monotectics, peritectics, eutectics. The description of microstructure evolution in such systems requires the introduction of multiple order parameters.

3. Multiple-Order Parameter Models

To describe equilibria of multiple phases like e.g. a liquid phase and two solid phases in eutectic solidification, an additional order parameter is required to distinguish the two solid phases (when liquid and solid are distinguished according to the fraction of solid introduced in chapter 1) . A possible choice is the interpretation of solute concentration as an order parameter as the two solids significantly differ in their composition [13].

Another approach is the definition of an order parameter ϕ_i for each of the phases i , which may be interpreted as the fraction of phase i within a certain control volume [14,15]. This interpretation naturally leads to the constraint:

$$\Sigma\phi_i = 1 \tag{3}$$

Besides assigning the different order parameters to different thermodynamic phases, these order parameters can moreover be used to distinguish different grains of a specific phase in order to treat phenomena like grain selection [16].

The evolution-equations for the different order parameters ϕ_i are obtained by functional minimization (- obeying constraint (3) by the use of Lagrange multipliers-) of a free energy functional essentially consisting of a summation of energy-contributions of all dual phase-boundaries i $\leftarrow$ $\rightarrow$ j . This kind of funtional allows for the individual treatment of the different interfaces with their specific properties like e.g. surface energy σ_{ij} or latent heat L_{ij} and eventually even with their own time scale [17]. Depending on the choice of the dual potential(s) (double obstacle or double well, see chapter 1) additional energy terms related to triple junctions may have to be considered [18].

It could be shown - both analytically (for the sharp interface limit) and numerically - that the respective equations of evolution lead to equilibria obeying classical laws for equilibria at triple junctions like e.g. Young's law describing the angles between different phases as a funtion of the different interfacial energies [19].

Simulations have meanwhile been performed on binary alloys like peritectic systems, e.g. high temperature superconductors [20] or steel [21,22], and eutectics [23] in both 2D and 3D using linearized binary phase diagrams. Especially the simulations of eutectic growth recovered well known analytical relations for the spacing between the two solid phases [24]. A review of the development of the phase-field concept for binary alloys can be found in [25].

4. Coupling with Thermodynamic Databases

Thermodynamic equilibrium data is requested for the evaluation of the local undercoolings and also to model diffusion. Because of the complexity of multiphase, multicomponent phase-diagrams, such data can best be evaluated using thermodynamic calculation software like ThermoCalc [26] or ChemSage [27]. However, on-line coupling by now is computationally too intensive to permit extended 2D/3D simulations on common workstations. In a simplified approach, thermodynamic calculations are performed as preprocessing for a given initial composition in order to linearize the liquidus (solvus) and solidus planes [28]. For simulations over a wide temperature range, a repeated update of the initial data during the calculation is advisable, when using this method.

Besides the free energies being deduced from thermodynamic databases also the adaptation of diffusion coefficients is necessary to simulate microstructure evolution in multicomponent systems. A suitable ansatz for implementation of diffusion coefficients including cross-diffusion is seen in the DICTRA code [29] to which phase-field computations have been compared recently [30].

5. Conclusions

The phase-field concept and its numerical realization seem to become a powerful tool for the prediction of microstructure evolution even in complex thermodynamic systems like real multicomponent multiphase alloys. Although already significant progress has been made within the last years many questions are still open in this rapidly growing field. These questions reach from fundamental issues about the origin of the potential being used to derive the phase-field equation over questions of coupling fields (e.g. electromagnetic) to the phase-field to issues of computational efficiency in terms of realistic 3D simulations in direct coupling with databases. Finally, the simulated microstructure has to be related to the macroscopic process parameters leading to its formation. First attempts to reach this objective are made by coupling macroscopic and microscopic simulations [31,32]. A real benefit for the production processes may result from simulations of specific initial conditions in a multiphase solidification problem, where for example a regular arrangement of precursor materials may lead to much more stable growth conditions [33,34] or to textured microstructures [35].

6. Acknowledgement

This paper in large parts represents a summary of ongoing activities and projects in ACCESS' microsimulation group, which are supported by British Steel, Hoogovens, Bayer as well as by BMBF and DFG. Thanks are due to the members of the group M.Apel , B.Böttger , H.J.Diepers, S.Gomes-Fries, U.Grafe, I.Steinbach and J.Tiaden for giving insight into their most recent results.

7. References

For (animated) simulation results on the topics presented please refer to:
http://www.access.rwth-aachen.de/mikrosim/micress.html

1.) G.Fix, Phase field models for free boundary problems, in "Free boundary Problems" Vol. II, Ed. A.Fasano, M.Primicerio (Piman, Boston 1983)
2.) J.B.Collins, H.Levine, Diffusive interface model of diffusion-limited crystal growth, Phys. Rev. B, Vol. 31 No. 9 (1985) 6119-6122
3.) G. Caginalp, P. C. Fife, Phys. Rev. B 33 11 (1986)7792

4.) A.A.Wheeler, W.J.Boettinger, G.B.McFadden, Phase-field model for isothermal phase transitions in binary alloys, Phys. Rev. A, Vol. 45 No. 10 (1992) 7424-7439

5.) R. Kobayashi, Modeling and numerical simulations of dendritic crystal growth, Physica D 63 (1993) 410-423

6.) S.-L. Wang, R.F.Sekerka, A.A.Wheeler, B.T.Murray, S.R.Coriell, R.J.Braun,

7.) G.B.McFadden, Thermodynamically-consistent phase-field models for solidification, Physica D 69 (1993) 189-200

8.) T.Ihle, H.Müller-Krumbhaar, Fractal and compact growth morphologies in phase transitions with diffusion transport, Phys. Rev. E, vol. 49 No. 4 (1994) 2972-2991

9.) A.Karma, W. J.Rappel, Numerical Simulation of Three-Dimensional Dendritic Growth, Phys. Rev. Lett. Vol. 77 No. 19 (1996) 4050-4053

10.) C. Beckermann, H.-J. Diepers, I. Steinbach, A. Karma, X. Tong: Modeling Melt Convection in Phase-Field Simulations of Solidification. Journal of Computational Physics, to be published.

11.) H.-J. Diepers, C. Beckermann, I. Steinbach: Modeling of Convection-Influenced Coarsening of a Binary Alloy Mush Using the Phase-Field Method. Modelling of Casting, Welding and Advanced Solidification VIII. Ed. by B.G. Thomas, C. Beckermann, TMS 1998, pp 565-572.

12.) H.J. Diepers, C. Beckermann, I. Steinbach: A Phase-Field Method for Alloy Solidification with Convection. Proc. of the 4th Int. Conf. on Solidification Processing, Sheffield, 7.-10. July 1997. Ed. by J. Beeck, H. Jones, pp 426-430.

13.) B.Nestler, A.A.Wheeler "A Multiphase-field model of eutectic and peritectic alloys:numerical simulation of growth structures", Physica D (1999) in press

14.) I. Steinbach, F. Pezzolla, B. Nestler, M. Seeßelberg, R. Prieler, G.J. Schmitz, J.L.L.Rezende: A phase field concept for multiphase systems. Physica D 94(1996), pp 135-147.

15.) J. Tiaden, B. Nestler, H.J. Diepers, I. Steinbach: The Multiphase-Field Model with an Integrated Concept for Modelling Solute Diffusion. Physica D (1998)115, pp 73-86.

16.) M. Apel, I. Steinbach: Phase Field Modeling of the Growth of multicrystalline-Silicon from the Melt. POLYSE '98, Intern. Conference on Polycrystalline Semiconductors, Schwäbisch Gmünd, 13.-18-9.1998, to be published

17.) I. Steinbach, F. Pezzolla: A Generalized field method for multiphase transformation using interface fields. Physica D, to be published.

18.) H.Garcke, B.Nestler, B.Stoth "A multiphase concept: numerical simulation of moving phase boundaries and multiple junctions" , SIAM J.on Applied Mathematics (1999) in press

19.) H.Garcke, B.Nestler, B.Stoth:" On anisotropic order parameter models for multiphase systems and their sharp interface limit" Physica D 115 (1998)87

20.) G.J. Schmitz, B. Nestler: Simulation of phase transitions in multiphase systems: peritectic solidification of (RE)Ba2Cu3O7-x superconductors. Mat. Sci. and Eng. B53(1998), pp 23-27.

21.) J. Tiaden U. Grafe: A Phase-Field Model for Diffusion and Curvature Controlled Phase Transformations in Steels. Subm. to: PTM International Conference on Solid-Solid Phase Transformations '99, Kyoto, May 1999.

22.) J. Tiaden : "Phase-Field simulations of the peritectic solidification of Fe-C" J. Crystal Growth 198/199 (1999) 1275-1280

23.) M. Seeßelberg, J. Tiaden, G.J. Schmitz, I. Steinbach: Peritectic and Eutectic solidification: Simulations of the microstructure using the multi-phase-field method. Proc. of the 4th Int. Conf. on Solidification Processing, Sheffield, 7.-10. July 1997. Ed. by J. Beeck, H. Jones, pp 440-443.

24.) M. Seeßelberg, J. Tiaden: Simulations of Binary Eutectic Microstructures Using the Multi-Phase-Field Method. Proc. 8th Conf. on Modelling of Casting, Welding and Advanced Solidification San Diego, June 1998, pp 557-564.

25.) I. Steinbach, G.J. Schmitz: Direct numerical simulation of solidification structure using the phase field method. Proc. 8th Conf. on Modelling of Casting, Welding and Advanced Solidification San Diego, June 1998, pp 521-532.

26.) B.Sundman, B. Jansson, and J.O. Anderson: CALPHAD 9 Vol.2, 1985, pp. 153-190

27.) G.Eriksson, K. Hack., S. Petersen, ChemApp - A programmable thermodynamic calculation interface Proceedings Werkstoffwoche '96, Symposium 8: Simulation, Modellierung, Informationssysteme,(1997) p.47 published by: DGM Frankfurt

28.) B. Böttger, U. Grafe, D. Ma, S.G. Fries. Application of Thermodynamic and Mobility Data to Two-Dimensional Modelling of Dendritic Growth During Directional Solidification. CALPHAD XXVIII, International Conference on Phase Diagram Calculations and Applications, Grenoble, Spring 1999, accepted.

29.) A.Engström et al: Met.Trans 25A (1994) 1127-1134

30.) U. Grafe, B. Böttger, J. Tiaden, S.G. Fries: Coupling of Multicomponent Thermodynamic Databases to a Phase Field Model: Application to Gamma Prime Growth in a Ternary NI-AL-CR Model Superalloy. Subm. to: PTM International Conference on Solid-Solid Phase Transformations '99, Kyoto, May 1999.

31.) G. Laschet, H.-J. Diepers, R. Prieler: Micro-Macro Simulation of a Laser Remelting Process. Subm to: Fifth U.S. National Congress on Computional Mechnanics, Boulder, Colorado 4.-6.8.99.

32.) G. Laschet, H.-J. Diepers, I. Steinbach: Micro-Macrosimulation of laser remelting of an aluminium coating on steel. Proc. of ECLAT '98. European Conference on Laser Treatment of Materials, Hannover 22.-23.09.1998, pp 265-270

33.) Ch. Wolters, J. Laakmann, S. Rex, G.J. Schmitz: Numerical simulation of the influence of Y2BaCuO5 particles on the growth morphology of peritectically solidifying YBa2Cu3O7-x. Proc EUCAS '93 Göttingen, ed.. H.C. Freyhardt, Oberursel: DGM 1993, p 353.

34.) G.J. Schmitz, O. Kugeler: Isothermal production of uniaxially textured YBCO superconductors using constitutional gradients. Physica C 275(1997), pp 205-210.

35.) G.J. Schmitz, A. Tigges, J.C. Schmidt: Texturing of (RE)Ba2Cu3O7-x thick films by geometrical arrangements of reactive precursors.
Supercond. Sci. Technol. 11(1998) pp 950-953.

Structure- Properties Relationships For Manganese Perovskites

B. Dabrowski, X. Xiong, O. Chmaissem, Z. Bukowski,

Department of Physics, Northern Illinois University, DeKalb, IL 60115

J.D. Jorgensen,

Materials Science Division, Argonne National Laboratory, Argonne, IL 60439

ABSTRACT

By combining the results of dc magnetization, ac susceptibility, magnetoresistivity, magnetostriction, and x-ray and neutron powder diffraction data for stoichiometric $La_{1-x}Sr_xMnO_3$ we have constructed a phase diagram that describes the magnetic, transport, and structural properties and the relationships among them as a function of composition and temperature. Correlations among physical and structural properties have been observed that are consistent with a competition between ferromagnetism and JT distortion. A metallic state occurs below the Curie temperature when both coherent and incoherent JT distortions are suppressed.

INTRODUCTION

For several decades the $La_{1-x}A_xMnO_3$ (A = Sr, Ba, Ca) compounds have been considered a classical example of materials for which the interactions between electrons are mediated mostly by magnetic super- and double-exchange interactions.[1-3] Recently, increased interest has been generated for these compounds because of a colossal magnetoresistive effect for the ferromagnetic phase ($x \geq 0.1$) near the Curie temperature, $T_C \sim 140 - 380$ K.[4] Renewed theoretical work has shown that the explanation of the transport and magnetic properties requires consideration of electron-phonon couplings in addition to magnetic interactions.[5-7] For $LaMnO_3$, a high-spin electronic configuration $t_{2g}^3 e_g^1$ for the Mn^{3+} ions with core-like t_{2g} states and extended e_g orbitals is susceptible to a strong electron-phonon coupling of the Jahn-Teller (JT) type that splits the e_g states into filled d_{z^2} and empty $d_{x^2-y^2}$ orbitals, and, thus, produces large asymmetric oxygen displacements around the Mn ions. The super-exchange interaction that induces ferromagnetic coupling within the Mn-O planes between JT-

27

R. Cloots et al. (eds.), Supermaterials, 27–36.

ordered orbitals of dissimilar symmetry and antiferromagnetic coupling perpendicular to the planes between orbitals of the same symmetry produces A-type, 3-dimensional antiferromagnetic structure. With substitution of divalent elements for La, electrons are removed from the d_{z^2} orbital causing weakening of both the JT-electron-phonon coupling and the superexchange interaction. Empty states in the e_g orbitals allow for the "breathing-mode" electron-phonon coupling of the charge ordered type that can produce charge localization and a double-exchange ferromagnetic interaction that favors metallic state.[3,8,9]

Competition between magnetic and electron-phonon interactions and structural transitions results in a very rich structure-property phase diagram for substituted $LaMnO_3$ materials.[10-12] Upon substitution of Sr, $La_{1-x}Sr_xMnO_3$ transforms from an antiferromagnetic insulator to a ferromagnetic insulator at $x \approx 0.1$ and to a ferromagnetic metal at $x \approx 0.16$. The Curie temperature increases from ~ 150 K for $x \approx 0.1$ to ~ 300 K for $x \approx 0.2$, and to ~380 K for $x \approx 0.3$. Three crystallographic phases have been identified at room temperature. For pure and lightly substituted materials, the orthorhombic Pbnm structure (O'), characterized by large coherent orbital ordering of the JT- type was found.[10,11,13-16] Above $x \approx 0.08 - 0.12$, a phase (O*) with the same orthorhombic Pbnm structure but characterized by a considerably smaller coherent JT-orbital ordering is observed.[10,11] At higher Sr substitution level, $x \approx 0.16 - 0.18$, the rhombohedral R3m structure (R), characterized by the absence of a coherent JT orbital ordering, was observed.[10]

The reported values of the compositions and temperatures for the structural, magnetic and resistive transitions vary substantially in the literature, depending on the sample processing method and synthesis conditions.[17] The discrepancies among existing phase diagrams may result from comparing behavior of compositions that are incorrectly assumed to have the same doping level.[18,19] We have recently shown that there can be large deviations from nominal stoichiometry for low and moderate substitution levels ($0 \leq x \leq 0.3$) - the perovskite structure can form a large concentration of vacancies, v, on both the La and Mn sites during synthesis under oxidizing conditions. Since the physical and structural properties are complex and are very sensitive to the hole concentration, we have established the intrinsic phase diagram (see Fig. 4) as a function of the hole doping; i.e., as a function of the Sr substitution level for samples that contain no metal-site vacancies ($v \approx 0$). [20-22]

SAMPLE PREPARATION AND CHARACTERIZATION

Polycrystalline samples of $La_{1-x}Sr_xMnO_3$ with $0.1 \leq x \leq 0.2$ and $\Delta x = 0.005$, were synthesized using a wet-chemistry method that leads to homogenous mixing of the metal ions. The final firing temperatures and oxygen atmosphere as well as holding times and cooling rates for obtaining stoichiometric samples were determined from TGA measurements performed using slow heating (2 deg/ min.) and cooling (0.6 deg/ min.) rates. Figure 1 shows the oxygen contents from TGA data for a sample with $x = 0.185$. The materials obtained after slow cooling in 100 and 20 % O_2 have the effective oxygen contents substantially above 3. Using the measured oxygen contents shown on Fig.1, the vacancy concentrations of these samples can be estimated (assuming equal amounts on both metal sites) as $v \approx d/(d+3)$. The weights of the samples approached stable levels, indicating a stoichiometric oxygen content of 3.002 ± 0.002 for temperatures around 1360 C. Stoichiometric $La_{1-x}Sr_xMnO_3$ samples were prepared in air by quenching from high temperatures, 1450 C for $x = 0.10$, and gradually lower temperatures for samples with larger x, to 1350 C for $x = 0.20$. Powder x-ray diffraction at room temperature confirmed single-phase material in each case and was used to determine structural phase boundaries. The $x = 0.10$ sample is orthorhombic-O' and a transition from the orthorhombic O' to the O* structure occurs around $x = 0.11$. The $x = 0.115$ composition is mixed phase, mostly O*. Compositions from $x = 0.12$ to 0.165 are pure orthorhombic O*. A transition from the orthorhombic O* phase to rhombohedral phase occurs for the $x = 0.17$ composition. Above $x = 0.17$, samples are rhombohedral.

PROPERTIES OF $La_{0.835}Sr_{0.165}MnO_3$

We have constructed the structure-properties phase diagram by combining the results of physical property and structural measurements. The magnetic and structural phase transitions were found to exhibit distinctive signatures in each kind of measurement. In Fig. 2 is shown magnetization and resistivity data for an $x = 0.165$ sample for which the ferromagnetic transition appears in the orthorhombic O* structure. Two features are seen in the magnetic data (Fig. 2a): The clearly visible ferromagnetic

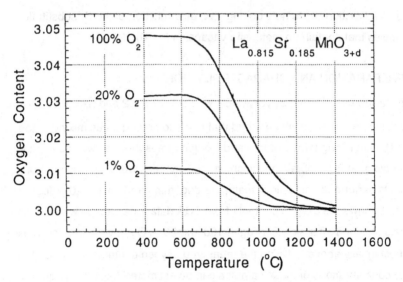

Fig. 1 Oxygen contents in 1, 20, and 100 % O_2 from thermogravimetric measurements for $La_{0.835}Sr_{0.165}MnO_{3+d}$ sample. The data were obtained during slow cooling at 0.6 deg/min.

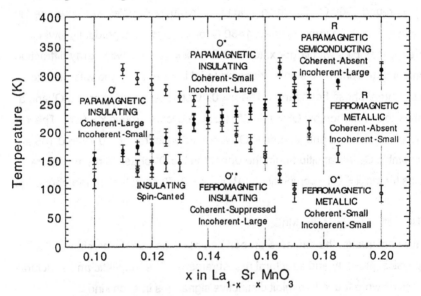

Fig. 4 Phase diagram summarizing the magnetic and structural properties of $La_{1-x}Sr_xMnO_3$ as a function of composition, $0.10 \leq x \leq 0.2$, and temperature, $12 \leq T \leq 350$ K. O*, O', O'*, and R stand for the small coherent JT-distorted orthorhombic ($\sigma_{JT} \sim 0.004$ Å), large coherent JT-distorted orthorhombic ($\sigma_{JT} \sim 0.05$ Å), suppressed coherent JT-distorted orthorhombic ($\sigma_{JT} \sim 0.02$ Å), and the rhombohedral phases, respectively.

transition at T_c = 256 K and the less apparent decreased-magnetization anomaly at T_s = 125 K. Both transitions can be precisely defined for small magnetic fields (~ 20 Oe) from the largest slope of the M(T) curves. These magnetic transitions are clearly correlated with the resistive transitions (Fig. 2b), the sharp drop of resistivity at T_c, and the metal-insulator transition at T_s. In addition, the resistivity and ac susceptibility have shown anomalies at T_r = 310 K that correspond to the rhombohedral-orthorhombic (R-O*) structural transition. Magnetization and resistivity data obtained for other $La_{1-x}Sr_xMnO_3$ samples showed similar transitions at consistently changing temperatures allowing the

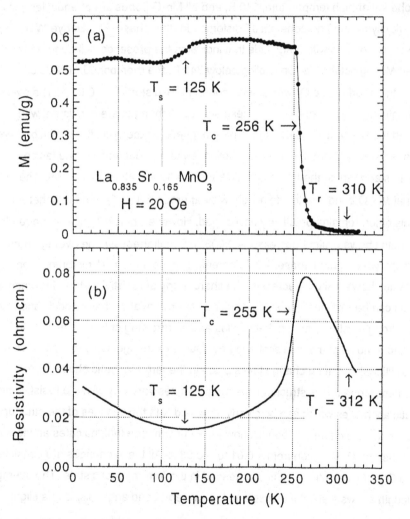

Fig. 2   Magnetization in a 20 Oe applied field (a) and resistivity (b) and for the $La_{0.835}Sr_{0.165}MnO_3$ sample.

correlated transport, magnetic, and structural phenomena to be characterized for each kind of phase transition that occurs in this system.

The neutron diffraction technique was used to accurately determine bond lengths, bond angles, and incoherent oxygen displacements. The pattern of the coherent displacements and the symmetry of the incoherent displacements have been used to obtain information about polaronic electron-phonon interactions in both the paramagnetic and ferromagnetic phases. Figure 3 shows the Mn-O bond lengths and the anisotropic Debye-Waller parameters for $La_{0.835}Sr_{0.165}MnO_3$. The sample is rhombohedral at high temperature, 340 K, and all Mn-O bonds are of equal length; i.e., there is (by symmetry) no coherent JT distortion for this phase. The Debye-Waller parameter is isotropic with magnitude that indicates the presence of incoherently distorted Mn-O_6 octahedra. On cooling below 340 K, the rhombohedral structure changes to orthorhombic O* with a small coherent JT distortion, ~ 0.01 Å. The average bond length increases significantly. These structural changes are consistent with increased resistivity at T ~ 312 K in the paramagnetic phase upon the transition from the rhombohedral to O* phase. On further cooling, an abrupt decrease of the lattice parameters appears at the ferromagnetic transition temperature T_c ~ 256 K. The individual Mn–O bond lengths do not show clear drops below T_c and the coherent Jahn-Teller distortion remains practically unchanged. However, below T_c, the average Mn-O bond length shows a slight decrease, ~ 0.003 Å. Simultaneously, the average bond angle displays a slight increase, ~ 0.4 degrees. The incoherent JT distortion along the Mn-O bond shows a slight decrease. The sharp drops of resistivity at the ferromagnetic transition can be explained, thus, as resulting from removal of charge localization below T_c by a strong electron-phonon JT-coupling and a narrowing of the band-width in addition to removal of the spin scattering by double-exchange. On cooling below ~ 160 K, an additional structural change takes place in the ferromagnetic phase. The temperature range of this structural change correlates very well with the resistive and magnetic anomalies which can be clearly identified with the vestiges of the structural O* to O' change. The equatorial bonds show an abrupt change that indicates an increase of the coherent JT distortion from ~ 0.01 to ~ 0.03 Å. All these changes are consistent with the O* to O' transition that is suppressed in the ferromagnetic state. The average bond-length shows a slight increase and the average bond angle displays a slight

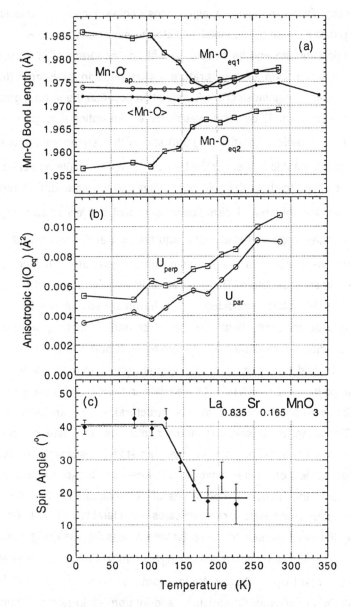

Fig. 3   Refined structural and magnetic parameters from neutron powder diffraction measurements as a function of temperature for $La_{0.835}Sr_{0.165}MnO_3$ sample: (a) Mn-O bond lengths and their average value, (b) anisotropic Debye-Waller factors perpendicular and parallel to Mn-O bond, and (c) spin angle with respect to crystallographic c-axis.

decrease on cooling. These changes are, again, consistent with the more insulating bonds in the O' phase as seen by the metal-insulator resistive transition.

The magnetic structure and its correlation with the coherent JT distortion have been studied as a function of temperature using neutron powder diffraction, magnetoresistance, and magnetostriction. A correlation between the ferromagnetic spin orientation and the coherent JT distortion was observed, indicating a coupling between the structural and magnetic properties. Fig.3(c) shows the refined spin angle ϕ with respect to the crystallographic c-axis. For the $La_{0.835}Sr_{0.165}MnO_3$ sample one can observe simultaneous transitions for both the spin orientation and the coherent JT distortion. Above 160 K, where the coherent JT distortion is very small, the refined spin angle, ϕ, is ~ 18°. The spin angle jumps to 40° at 120 K and remains at this value to the lowest temperature, while the JT distortion reaches its maximum value at ~ 120 K and remains constant to the lowest temperature. The magnetoresistance data have shown that the corresponding metal-insulator transition in the ferromagnetic phase is practically insensitive to applied magnetic field. On the other hand, the temperature of the ferromagnetic transition temperature increases from 256 K at 0 T to 290 K at 7 T. With increasing magnetic field the ferromagnetic transition begins to overlap with the resistive anomaly originating from the structural R/O* transition in the paramagnetic phase. As a result the resistive anomaly caused by the structural R/O* transition is clearly suppressed. The magnetostriction measurements are consistent with these observations. For x = 0.165, the pronounced minimum at 250 K is clearly observed only at 0 T. At a magnetic field of 5 T the minimum is reduced and broadened and at 10 T it is not observable at the highest temperatures of our measurements (300 K). The observed large change of magnetostriction indicates considerable changes of the lattice constants during the ferromagnetic transition for the $La_{0.835}Sr_{0.165}MnO_3$ sample with the O* structure as seen from neutron diffraction.[20] Similar rapid decreases of lattice constants and anisotropic Debay-Waller factors were observed for the $La_{0.75}Ca_{0.25}MnO_3$ sample with O* structure, and interpreted in terms of removal of the incoherent JT-distortions during the ferromagnetic transition.[23] The broad maximum of magnetoresistance was observed in the temperature range 120 - 135 K to be practically independent of magnetic field. Thus, it appears that when the structural

O*/O' transition occurs in the ferromagnetic phase, no additional structural changes are induced by the external magnetic fields up to 10 T.

PHASE DIAGRAM

Based on similar data obtained for the whole series of samples, the following structure-properties relationships were developed for moderately substituted $La_{1-x}Sr_xMnO_3$. In the high temperature paramagnetic regime three structural phases are present: Orthorhombic Pbnm structure characterized by the large coherent orbital ordering of the JT-type and small incoherent JT-distortions, Orthorhombic Pbnm structure characterized by a small coherent JT-orbital ordering and large incoherent JT-distortions, and Rhombohedral R3m structure characterized by the absence of a coherent JT-orbital ordering and large incoherent JT-distortions. For $0.10 < x < 0.145$, during the ferromagnetic transition, the orthorhombic O' phase changes to an orthorhombic O'* phase characterized by suppressed coherent JT-orbital ordering and large incoherent JT-distortions. For $0.15 < x < 0.17$, on cooling during the ferromagnetic transition the insulating orthorhombic O* phase changes to a metallic O* phase characterized by small coherent and small incoherent JT-distortions. Below the Curie temperature, in the ferromagnetic phase, this orthorhombic O* phase changes to an orthorhombic O'* phase characterized by suppressed coherent JT-orbital ordering and large incoherent JT-distortions that is insulating. For $0.175 < x < 0.20$, on cooling during the ferromagnetic transition the semiconducting R phase changes to a metallic R phase with small incoherent JT-distortions. Below the Curie temperature, in the ferromagnetic phase, this R structure changes to an orthorhombic Pbnm structure O* characterized by small coherent JT-orbital ordering and small incoherent JT-distortions that is still metallic. Development of ferromagnetic order, thus, suppresses the coherent JT-distortions for the O' phase and incoherent distortions for the O* and R phases. The arrangement of the spins varies from ferromagnetic ordered mainly along the b-axis ($x = 0.11$) to ferromagnetic almost along the c-axis ($x = 0.185$) in the Pbnm symmetry.

The data clearly show that the ferromagnetic phase has lower resistivity than the paramagnetic phase. However, the ferromagnetic transition is accompanied by structural changes that can also cause a decrease of resistivity. The suppression of coherent JT-distortions lowers the resistivity but is not sufficient to produce a metallic phase unless large incoherent JT-distortions are also removed. The electron-phonon Jahn-Teller interactions and the magnetic super-exchange and double-exchange interactions are of similar strength. These interactions compete for the same e_g electrons and, thus, lead to colossal spin-charge-lattice coupling.

ACKNOWLEDGMENT

The work at NIU was supported by the ARPA/ONR and the State of Illinois under HECA, and at ANL by US DOE (W-31-109-ENG-38)

REFERENCES

1. C. Zener, Phys. Rev. 82, 403 (1951).
2. P.W. Anderson and H. Hasegawa, Phys. Rev. 100, 675 (1955).
3. J.B. Goodenough, *Magnetism and chemical bond*, Wiley, New York, 1963.
4. R.M. Kusters, J. Singelton, D.A. Keen, R. McGreevy, and W. Hayes, Physica B 155, 362 (1989).
5. A.J. Millis, P.B. Littlewood, and B.I. Shraiman, Phys. Rev. Lett. 74, 5144 (1995).
6. A.J. Millis, B.I. Shraiman, and R. Mueller, Phys. Rev. Lett. 77, 175 (1996).
7. H. Roder, J. Zhang, and A.R. Bishop, Phys. Rev. Lett. 76, 1356 (1996).
8. P.A. Cox, *Transition Metal Oxides*, Clarendon Press, Oxford, 1995.
9. G.H. Jonker and J.H. Van Santen, Physica 16, 337 (1950).
10. A. Urushibara, Y. Moritomo, T. Arima, A. Asamitsu, G. Kido, Y. Tokura, Phys. Rev. B 51, 14103 (1995).
11. H. Kawano, R. Kajimoto, and H. Yoshizawa, Phys. Rev. B 53, R14709 (1996).
12. J.-S. Zhou, J.B. Goodenough, A. Asamitsu, and Y. Tokura, Phys. Rev. Lett. 79, 3234 (1997).
13. E.O. Wollan and W.C. Koehler, Phys. Rev. 100, 545 (1955).
14. J.B.A.A. Elemans, B. Van. Laar, K.R. Van Der Veen, and B.O. Loopstra, J. Solid State Chem. 3, 238 (1971).
15. Q. Huang, A. Santoro, J.W. Lynn, R.W. Erwin, J.A. Borchers, J.L. Peng, and R.L. Greene, Phys. Rev. B 55, 14987 (1997).
16. J. Rodriguez-Carvajal, M. Hennion, F. Moussa, A.H. Moudden, L. Pinsard, and A. Revcolevschi, Phys. Rev. B 57, R3189 (1998).
17. J.F. Mitchell, D.N. Argyriou, C.D. Potter, D.G. Hinks, J.D. Jorgensen, and S.D. Bader, Phys.Rev. B 54, 6172 (1996).
18. J.A.M. Van Roosmalen, and E.H.P. Cordfunke, J. Solid State Chem. 110, 106 (1994).
19. B. Dabrowski, K. Rogacki, X. Xiong, P.W. Klamut, R. Dybzinski, J. Shaffer, and J.D. Jorgensen, Phys. Rev. B 58, 2716 (1998).
20. B. Dabrowski, X. Xiong, Z. Bukowski, R. Dybzinski, P.W. Klamut, J.E. Siewenie, O. Chmaissem, J. Shaffer, C.W. Kimball, J.D. Jorgensen, Phys. Rev. B
21. X. Xiong, B. Dabrowski, O. Chmaissem, Z. Bukowski, S. Kolesnik, R. Dybzinski, C.W. Kimball, J.D. Jorgensen Phys. Rev. B
22. B. Dabrowski, L. Gladczuk, A. Wisniewski, Z. Bukowski, R. Dybzinski, A. Szewczyk, M. Gutowska, S. Kolesnik, C.W. Kimball, and H. Szymczak (unpublished)
23. P.G. Radaelli, G. Iannone, M. Marezio, H.Y. Hwang, S-W. Cheong, J.D. Jorgensen and D.N. Argyriou, Phys. Rev. B 56, 8265 (1997).

VIBRONIC PHENOMENA AT LOCALIZED-ITINERANT AND MOTT-HUBBARD TRANSITIONS

J.B. GOODENOUGH AND J.-S. ZHOU
Texas Materials Institute, ETC 9.102
University of Texas at Austin, Austin, TX 78712-1063

ABSTRACT

It is argued from the pressure dependence of the transport properties that the Mott-Hubbard transition in the AMO_3 perovskites is first-order with a dynamic phase segregation appearing in the metallic phase on the approach to the transition from the itinerant-electron side. A transition with decreasing bandwidth from a strongly enhanced Pauli paramagnetism toward a Curie-Wiess law is observed in the metallic CuO_3 array of $La_{1-x}Nd_xCuO_3$, $0 \leq x \leq 0.6$, and in the metallic phase of the compounds $LnNiO_3$, $Ln = La$, Pr, Nd, $Sm_{0.5}Nd_{0.5}$. The $LnNiO_3$ family undergoes an antiferromagnetic-insulator to metal transition at a temperature T_t that is sensitive to $^{18}O/^{16}O$ isotope exchange and disappears in $LaNiO_3$ and $PrNiO_3$ under 15 kbar hydrostatic pressure. We suggest a bandwidth of the form

$$W = W_b \exp (-\lambda \varepsilon_{sc}/h\omega_0)$$

where $\lambda \sim \varepsilon_{sc}/W_b$, W_b is the tight-binding bandwidth, and a pressure-sensitive ω_0^{-1} is the period of the locally cooperative oxygen displacements that define strong-correlation fluctuations stabilized by an energy ε_{sc}. $LaMnO_3$ exhibits an insulator-conductive electronic transition at a cooperative Jahn-Teller orbital ordering below T_{JT}; the magnetic susceptibility obeys a Curie-Weiss law in which θ, but not C, changes discontinuously at T_{JT}. We propose a double-exchange coupling involving vibrons.

INTRODUCTION

The transition from localized to itinerant electronic behavior in a single-valent compound is different from that in a mixed-valent compound, but in each case unusual electron lattice interactions may occur.

In a single-valent compound, the Mott-Hubbard transition that occurs at a critical ratio W/U of the bandwidth W to the on-site electron-electron coulomb energy U is between weakly and strongly correlated itinerant electrons; localization of the electrons occurs at a smaller W/U ratio. The Mott-Hubbard transition is generally assumed to be a smooth and globally homogeneous transition; but this assumption has not been tested systematically by experiment, and failure of theory to include strong electron-lattice interactions at cross-over motivates such a test.

R. Cloots et al. (eds.), Supermaterials, 37–47.
© 2000 *Kluwer Academic Publishers. Printed in the Netherlands.*

The suggestion that strong electron-lattice interactions should be present at cross-over comes from the virial theorem for central-force fields,

$$2 <T> + <V> = 0 \qquad (1)$$

which states that a discontinuous increase in the volume occupied by an electron on crossing a transition results in a discontinuous decrease in the mean kinetic energy $<T>$ of the electrons and therefore a discontinuous decrease in the magnitude $|<V>|$ of the mean potential energy. For antibonding electrons in an AMO_3 perovskite, a discontinuous decrease in $|<V>|$ translates to a discontinuous decrease in the equilibrium (M-O) band length; the transition is, therefore, first-order with a longer equilibrium (M-O) bond length for the phase with more localized electrons:

$$(M-O)_{loc} > (M-O)_{itin} \qquad (2)$$

Early studies of the evolution of physical properties across the Mott-Hubbard transition were frustrated by phase segregation or by stabilization of a static charge-density wave (CDW) [1]. However, the perovskite structure allows experimental study of systems that approach the Mott-Hubbard transition from either the itinerant-electron or the localized-electron side, and recent experiments indicate that strong electron-lattice interactions may introduce dynamic phase segregations and/or stabilization of vibronic states in which electronic and phonic states are hybridized. Neither dynamic phase segregation nor vibronic states are detectable with conventional diffraction experiments: faster probes are needed for direct detection; but indirect signatures are useful, particularly if characterized in conjunction with a direct probe.

In a mixed-valent system, the transition from small-polaron to itinerant-electron behavior occurs at a critical ratio τ_h/ω_0^{-1} of the electron-transfer time τ_h between neighboring like atoms to the period ω_0^{-1} of the optical-mode atomic displacements that define the volume of the polaron. Dynamic phase segregation and/or formation of vibronic states may also be found in mixed-valent systems on crossing the critical ratio $\tau_h/\omega_0^{-1} \approx 1$. For example, the intrinsic colossal magnetoresistance (CMR) phenomenon appearing in the manganese-oxide perovskites with $Mn(IV)/Mn = 0.3$ illustrate this situation, as has been discussed elsewhere [2]. The high-temperature superconductivity of the mixed-valent copper oxides appears to provide another illustration [3].

In this paper, we report how the magnetic and transport properties change with W on the approach to the Mott-Hubbard transition in some single-valent AMO_3 perovskites. The bandwidth W was changed in two ways: by isovalent substitution of smaller A cations or by hydrostatic pressure. The study was motivated by three questions: (1) Are the Mott-Hubbard transitions homogeneous and smooth or first-order and heterogeneous (2) Can a heterogeneous electronic system condense into a static CDW below an order-disorder transition temperature T_t? (3) What is the role of electron-lattice interactions?

$Sr_{1-x}Ca_xVO_3$

The perovskite system $Sr_{1-x}Ca_xVO_3$ contains a partially filled π^* band that decreases in width with increasing x [1]. Over the entire compositional range $0 \leq x \leq 1$, the resistivity $\rho(T)$ of $Sr_{1-x}Ca_xVO_3$ has a metallic temperature dependence and exhibits an enhanced Pauli paramagnetism [4,5]. However, $\rho(T)$ is too high to be interpreted with a conventional band model and $CaVO_3$ becomes antiferromagnetic if it is only slightly oxygen-deficient. With increasing x, the system clearly approaches the Mott-Hubbard transition from the itinerant-electron side. Ultraviolet photoemission spectra (PES) from the University of Tokyo [6,7] have shown the coexistence of Fermi-liquid electrons and strongly correlated electrons in a lower Hubbard band with a continuous shift of spectral weight from the Fermi-liquid to the lower Hubbard band with increasing x and a maximum in the effective mass m^* of the Fermi-liquid states at an intermediate value of x. Since diffraction data indicate a homogeneous single phase for all x, the PES data suggested to us the existence of strong-correlation fluctuations in a Fermi-liquid sea. The Brinkman-Rice (B-R) theory of mass enhancement on the approach to the Mott-Hubbard transition gives a ratio

$$m^*/m_e = [1 - (U/U_C)^2]^{-1} \tag{3}$$

and we proposed [8] that a singularity in m^* at $U = U_C$ is avoided by a first-order transition; as a result, a dynamic phase segregation is associated with locally cooperative oxygen displacements that define strong-correlation fluctuations in volumes of longer $(V-O)_{loc}$ bond lengths. In order to test this interpretation without the complication of two different A-site cations, we measured the resistivity $\rho(T)$ and thermoelectric power $\alpha(T)$ of $CaVO_3$ under different hydrostatic pressures [8]. An increase with pressure in the magnitude of the room-temperature thermoelectric power, $d|\alpha(300\ K)|/dP > 0$, confirmed that m^* increases with W in this compound. Moreover, a suppression of the low-temperature phonon-drag component of $\alpha(T)$ was partially lifted with increasing pressure, which is consistent with a decrease in the density of strong-correlation fluctuations with increasing W. These measurements not only confirmed the more direct PES evidence for strong-correlation fluctuations in a Fermi-liquid sea; they also provided an indirect signature in the pressure dependence of $\alpha(T)$ for a heterogeneous vs. homogeneous character of an electron system.

In a single-valent AMO_3 perovskite, the presence of disordered strong-correlation fluctuations would introduce strong electron-lattice interactions; the locally cooperative oxygen displacements that define a fluctuation would have a period ω_0^{-1} that is too short to be visible with a diffraction experiment. By analogy with the band narrowing due to polaron formation [9], the fluctuations would narrow the bandwidth to

$$W = W_b \exp(-\lambda\varepsilon_{sc}/\hbar\omega_0) \tag{4}$$

where W_b is the tight-binding bandwidth, ε_{sc} is the stabilization energy of a strong-correlation fluctuation, and $\lambda \sim \varepsilon_{sc}/W_b$ is a screening parameter. Pressure would not only increase W_b; it would also increase ω_0, which would have a more profound influence on a $dW/dP > 0$. Moreover, ^{18}O for ^{16}O isotope exchange would decrease ω_0, thus favoring the phase with more localized electrons. The combination of pressure and isotope exchange can be used to great effect, as has been demonstrated [10] for a mixed-valent manganese oxide where strong electron-lattice interactions introduce an expression for W similar to that of Equation (4).

$La_{1-x}Nd_xCuO_3$

Oxygen-stoichiometric $La_{1-x}Nd_xCuO_3$ was prepared over the range $0 \le x \le 0.6$ in a belt apparatus at 25 kbar pressure and $900 - 1000°C$ [11]. Four-probe resistivity data $\rho(T)$ taken on rectangular samples cut from the central portion of the pressed pellets showed a metallic temperature dependence that was lower by a factor 10^{-4} than what had been reported in the

40

literature for $LaCuO_3$ [12]. In this system, the CuO_3 array contains a half-filled σ^* band, and the thermoelectric power was correspondingly small. However, a suppression of the low-temperature phonon-drag component as well as a $d|\alpha(300\ K)|/dP > 0$ were characteristic of a heterogeneous electronic system with a band that narrows with increasing x; but the Mott-Hubbard transition is not quite reached as in the $Sr_{1-x}Ca_xVO_3$ system.

Fig. 1 shows the evolution with increasing Nd concentration x of the paramagnetic susceptibility $\chi(T)$ of the CuO_3 array. To subtract the Nd^{3+}-ion contribution to $\chi(T)$, we used $NdAlO_3$, which had a measured Curie constant C = 1.615 close to the free-ion value C = 1.617 and a large, negative Weiss constant θ = -95 K. Taking proper account of θ proved critical. The $\chi(T)$ curve for $LaCuO_3$ can be seen to be temperature-independent above 200 K, but it is enhanced by a factor 10^2 relative to a conventional Pauli paramagnetism. Below 200 K, the temperature dependence of $\chi(T)$ cannot be attributed to magnetic impurities since their contribution would only appear at extremely low temperatures given the high value of $\chi(T)$ above 200 K. At 300 K, the magnitude of $\chi(T)$ remains higher than the enhanced Pauli paramagnetism, so a strong temperature dependence persists over the temperature range of measurement. However, the Curie constant remains too large for a localized-electron Curie-Weiss paramagnetism.

The susceptibility $\chi(300\ K)$ of the CuO_3 array increases linearly with x by more than a factor 10 in the range $0 \leq x \leq 0.6$. This extraordinary change is due to a narrowing of the width W of the σ^* band of the CuO_3 array at the approach to the Mott-Hubbard transition from the itinerant-electron side.

With increasing x, the tight-binding bandwidth $W_b \sim \varepsilon_\sigma \lambda_\sigma^2 \cos\phi$ is narrowed by two factors: (1) perturbation of the periodic potential by two different A cations and (2) an increase in the mean bending angle ϕ of the $(180° - \phi)$ Cu-O-Cu bond angles to accommodate the smaller Nd^{3+} ion.

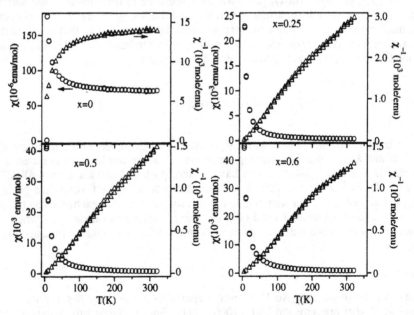

Figure 1 Magnetic susceptibility $\chi(T)$ and $\chi^{-1}(T)$ versus temperature for the CuO_3 array of $La_{1-x}Nd_xCuO_3$.

However, neither of these factors appears to be large enough to account for the remarkable change in χ (300 K). On the other hand, a change of ω_0 with ϕ could change the W of equation (4) significantly.

In order to introduce a T^{-1} temperature dependence into χ(T) for a homogeneous metallic band, it is necessary to have states at the Fermi energy ε_F with an $\varepsilon(k)$ dispersion less than kT. The Stoner enhancement does not introduce a flattening of $\varepsilon(k)$, but the B-R mass enhancement of equation (3) does flatten $\varepsilon(k)$ near ε_F. Mott [13] has pointed out that this flattening within an energy range $\Delta E \approx kT_d$ can introduce temperature-independent magnetic moments that give a 1/kT contribution to χ(T) below T_d. At $T > T_d$, the Pauli paramagnetic contribution from the more dispersive electronic states would dominate the susceptibility. This picture could apply to the χ(T) curve for LaCuO$_3$ with a $T_d \approx 200$ K. However, it is not obvious from qualitative considerations how to explain with this model the evolution of χ(T) with x, which shows little change in T_d and an increase of over a factor 10 in χ(300 K).

An alternative model is suggested by the transport data, which give indirect evidence of strong-correlation fluctuations coexisting with Fermi-liquid states. The fluctuations would impart a Curie-Weiss component to χ(T), and the fluctuation concentration would increase as W narrowed. The Curie-Weiss component from the strong-correlation fluctuations would give a linear increase in χ^{-1}(T) above T_d until saturation occurred on crossing the temperature-independent Pauli contribution.

LnNiO$_3$

The LnNiO$_3$ perovskites (Ln = rare-earth) contain low-spin Ni(III) with an orbitally twofold-degenerate σ^* band one-quarter filled and a tight-binding bandwidth $W_b \sim \varepsilon_\sigma \lambda_\sigma^2 \cos \phi$ as in LaCuO$_3$.

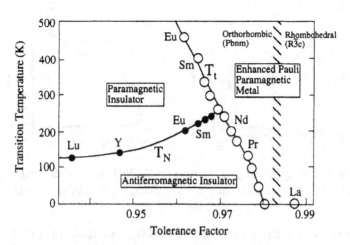

Figure 2 The insulator-metal transition temperature T_t and Néel temperature T_N for the LnNiO$_3$ family (adapted from [16]).

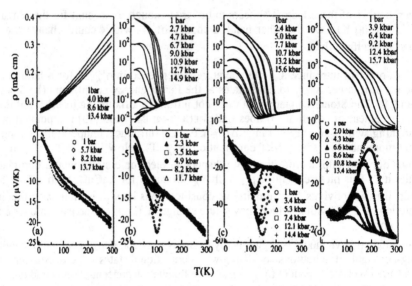

Figure 3 *The resistivity ρ(T) and thermoelectric power α(T) under different hydrostatic pressures for (a) LaNiO₃, (b) PrNiO₃, (c) NdNiO₃, and (d) Sm₀.₅Nd₀.₅NiO₃.*

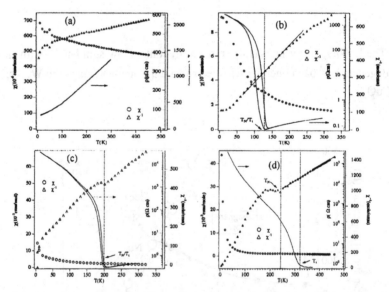

Figure 4 *The χ(T) and χ⁻¹(T) curves for the NiO₃ array of the samples of Fig. 3; the ρ(T) curves (solid lines) locate Tᵣ. (a) LaNiO₃, (b) PrNiO₃, (c) NdNiO₃, (d) Nd₀.₅Sm₀.₅NiO₃.*

Although the metallic character of rhombohedral $LaNiO_3$ was established in 1965 [14], it was not until 1991 that high-pressure synthesis of the orthorhombic samples $Ln_{1-x}Ln'_xNiO_3$ was achieved from $La_{1-x}Pr_xNiO_3$ to $EuNiO_3$ by Lacorre *et al* [15]. As indicated in Fig. 2, the orthorhombic samples exhibit an insulator-metal transition at a temperature T_t [9] and a transition to an unusual long-range antiferromagnetic order below a $T_N \leq T_t$ [17] that has been interpreted to represent a charge-density-wave/spin-density-wave (CDW/SDW) phase [11]. Fig. 3 shows the transport properties $\rho(T)$ and $\alpha(T)$ under different hydrostatic pressures for $LaNiO_3$, $PrNiO_3$, $NdNiO_3$, and $Sm_{0.5}Nd_{0.5}NiO_3$. The oxygen-stoichiometric (3.00 ± 0.01) samples were prepared under high oxygen pressure by B. Dabrowski. We call attention here to the following features:

(1) $LaNiO_3$ is a "bad metal" with a $\rho(T)$ too high to be accounted for with conventional band theory. Moreover, the $\alpha(T)$ curves show a suppression of the phonon-drag component, but an enhancement of both this component and m^* (by 15% in 14 kbar) under hydrostatic pressure. These are the characteristics of a heterogenous electronic system on the approach to a Mott-Hubbard transition from the itinerant-electron side. Although the transition at T_t in the orthorhombic samples prevents observation of the phonon-drag component of $\alpha(T)$, the "bad metal" $\rho(T)$ and a $d|\alpha(300 K)|/dP > 0$ indicate a similar situation in the metallic phase of the orthorhombic samples.

(2)   A first-order insulator-metal transition occurs at $T_N = T_t \approx 130$ K and 200 K, respectively, in $PrNiO_3$ and $NdNiO_3$; the transition in $PrNiO_3$ remains weakly first order, but decreases linearly with pressure to 85 K at 13 kbar. Remarkably, in this same pressure interval the low-temperature resistivity changes from a semiconductive to a metallic temperature dependence and $\rho(15 K)$ drops over five orders of magnitude. These changes are also reflected in the thermoelectric power $\alpha(T)$. At 13 kbar pressure, there appears to be little change in the number and mobility of the charge carriers in $PrNiO_3$ on crossing T_t. Comparison of $LaNiO_3$ at atmospheric pressure and $PrNiO_3$ under 14.9 kbar shows a significantly lower $\rho(T)$ and higher $|\alpha(T)|$ for the $PrNiO_3$ sample. Pressure does not appear to bring $PrNiO_3$ toward an orthorhombic-rhombohedral transition.

(3)   The T_N and T_t are separated in $Sm_{0.5}Nd_{0.5}NiO_3$ and the transition becomes second-order at T_t; it remains only weakly first-order at T_N. Although T_N and T_t are both lowered under pressure, the interval between them changes little, which is contrary to what would be expected from Fig. 2 if pressures simply increased W_b as has been assumed in the literature. In fact, comparisons of $PrNiO_3$ under 15 kbar and $LaNiO_3$ at atmospheric pressure, $NdNiO_3$ under 7.4 kbar and $PrNiO_3$ at atmospheric pressure, and of $Nd_{0.5}Sm_{0.5}NiO_3$ under 8.6 kbar and $NdNiO_3$ at atmospheric pressure show that the transport properties are quite dissimilar even though the transition temperatures T_t are the same for each pair. We conclude that pressures below 15 kbar have a relatively small influence on W_b, which means that if T_t occurs at a critical bandwidth W_c, the principal contribution to the large $dT_t/dP < 0$ is associated with a pressure dependence of ω_0 in Equation (4). It follows that the remarkable change with pressure in the number and/or mobility of the charge carriers in the CDW/SDW phase is also associated with a pressure dependence of ω_0. Pressure increases ω_0 and therefore destabilizes trapping of the fluctuations in a static CDW. Moreover, a second-order transition at $T_t > T_N$ indicates an order-disorder transition, presumably due to an ordering of the strong-correlation fluctuations into the CDW/SDW. Indeed, the magnetic order below T_N can be understood if the ferromagnetic Ni-O-Ni slabs contain Fermi-liquid states and these slabs are coupled antiferromagnetically by superexchange across an O^{2-} ion [18].

(4)   The observation [19] that substitution of ^{18}O for ^{16}O in $NdNiO_3$ increases $T_t = T_N$ by 10.3 K without introducing any change in W_b can be accounted for with a W described by equation (4).

44

Fig. 4 shows the evolution of χ(T) for the four compositions. The χ(T) curve for LaNiO$_3$ has a weak temperature dependence with too large a C and negative θ to be interpreted with a localized spin s = 1/2. As the size of the Ln^{3+} ion decreases, χ(T) of the NiO$_3$ array increases dramatically. Both C and θ become progressively smaller in the metallic phase of NdNiO$_3$ and Nd$_{0.5}$Sm$_{0.5}$NiO$_3$, but they remain too large for a localized spin s = 1/2. Significantly, there is no change in χ^{-1} versus T on crossing T$_t$ in Nd$_{0.5}$Sm$_{0.5}$NiO$_3$, which is compatible with an order-disorder transition of strong-correlation fluctuations that are present in the metallic phase; it is not compatible with a homogeneous transition driven by Fermi-surface nesting.

The recent report [20] of two distinguishable Ni sites in YNiO$_3$, one with longer and the other with shorter Ni-O bond lengths, may represent another type of order of the strong-correlation fluctuations; it does not require a charge disproportionation from one Ni to the other so much as a change in the extent of Ni:e-O:2p hybridization from one Ni to the other.

LaMnO$_3$

Finally, we turn briefly to LaMnO$_3$. Whereas the single e electron of octahedral-site, low-spin Ni(III):t^{6}e^1 occupies a quarter-filled σ* band of e-orbital parentage in LaNiO$_3$, the high-spin Mn(III):t^{3}e^1 configuration contains localized t^3 electrons of spin S = 3/2 that remove the spin degeneracy of the e orbitals. LaMnO$_3$ undergoes a cooperative Jahn-Teller orbital ordering below a T$_{JT}$ ≈ 750 K that removes the remaining twofold e-orbital degeneracy. Cooperative rotations of the MnO$_{6/2}$ octahedra about a cubic [110] axis accommodate a tolerance factor t < 1 and lower the symmetry to O-orthorhombic above T$_{JT}$. Superposition of the orbital ordering reduces the c/a ratio, and we designate the orbitally ordered phase O'-orthorhombic. Orbital ordering also introduces anisotropic superexchange interactions: they are ferromagnetic in the (001) planes and antiferromagnetic along the c-axis. [21].

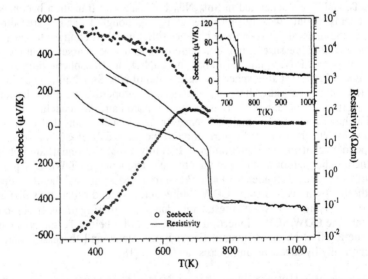

Figure 5 The resistivity ρ(T) and thermoelectric power α(T) of single-crystal LaMnO$_3$ obtained on the initial heating and cooling taken in 10^{-3} torr with I = 0.05 mA.

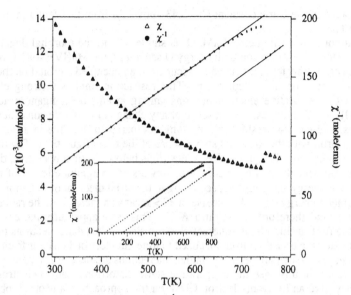

Figure 6   Magnetic susceptibility χ (T) and χ^{-1}(T) of single-crystal LaMnO$_3$; $\chi_{||}$ and $\chi \perp$ could not be resolved in the interval $T_N < T < T_{JT}$.

Fig. 5 shows atmospheric-pressure α(T) and four-probe ρ(T) data for a single crystal of LaMnO$_3$ grown in an IR-heated image furnace (NED SC-35MD) under a flow of pure Ar gas. Although these measurements and the χ (T) of Fig. 6 were carried out in a vacuum of 10^{-3} torr, an irreversible change in α(T) from – 600 μV/K to about + 550 μV/K at 300 K after heating to 1000 K reflects a small oxidation that takes place in the thermal cycle. The high magnitude of a reflects polaronic conduction with

$$\alpha = -(k/e)\ln[(1-c)/c] \tag{5}$$

which gives c $\approx$ 0.0009 in the reduced and the oxidized states.

The conductive state above T_{JT} is unusual. A temperature-independent α(T) is typical of polaronic behavior, but a temperature-independent ρ(T) is not, even for a motional enthalpy $\Delta H_m < kT$. Itinerant electrons having a constant mean free path would give a temperature-independent ρ(T), but the high value of ρ(T) would require a very low mobility $\mu = e_o\tau_s/m^*$ and therefore an exceptionally high value of the effective mass. Strong electron coupling to dynamic Jahn-Teller deformations would hybridize phonon and electron states to form *vibrons* with the required large m* and flatten the $\epsilon(\mathbf{k})$ curve so as to give a nearly temperature-independent α(T). Real charge transfer between Mn atoms in the conductive state would give a new type of Mn-O-Mn double-exchange spin-spin coupling involving vibrons. Zener double exchange requires a mixed valence; de Gennes double exchange might be extrapolated to a single-valent system, but his mechanism involves conventional itinerant electrons. A double-exchange coupling would be stronger than a superexchange interaction in which a virtual hopping of localized e electrons to empty e orbitals is correlated by the dynamic Jahn-Teller deformations. An increase in the strength of the interatomic interactions on crossing T_{JT} is signaled by an abrupt change in the Weiss constant θ from 52 K to 177 K with no change in the Curie constant C = 3.43 above T_{JT} from that found in the interval 300 < T < 500 K.

Although stoichiometric $LaMnO_3$ is single-valent, the e-orbital degeneracy lowers the on-site electron-electron coulomb energy U separating the Mn(IV)/Mn(III) redox energies provided electron transfer is restricted to occur from an occupied e orbital on the donor ion to an empty e orbital on the acceptor ion. This restriction implies a strong electron-lattice coupling with cooperative electron motions among groups of manganese atoms so as to retain Mn(III) configurations. The absence of any change in the Curie constant C on crossing T_{JT} implies an e-electron residence time long compared to the time to align the spin of a mobile e electron with the localized spin S = 3/2 of the t^3 configuration. The transition from cooperative, static ordering of localized e electrons below T_{JT} to cooperative, dynamic Jahn-Teller distortions of the Mn octahedral sites lowers the magnitude |<V>| of the e-electron mean potential energy and therefore, according to the virial theorem of Equation (1), decreases the mean kinetic energy <T>. A decrease in <T> means an increase in the radial extension of the e orbitals and therefore of the ratio W/U. This ratio apparently increases to near the critical value for transition to an itinerant-electron state. Thus the e electrons of $LaMnO_3$ are seen to approach the transition from localized to itinerant electron behavior from the localized-electron side; strong, dynamic electron-lattice interactions are once again manifest at the cross-over. However, in this case the cooperative static distortions lead to electron localization whereas they give an itinerant-electron CDW on the approach to a Mott-Hubbard transition from the itinerant-electron side.

We thank the NSF and, from Houston, Texas, the TCSUH and the Robert A. Welch Foundation for financial support.

REFERENCES

1. Goodenough, J.B. (1971) "Metallic Oxides", *Prog. Solid State Chem.* **5**, 145

2. Goodenough, J.B. (1999) "CMR in $Ln_{1-x}A_xMnO_3$ Perovskites," *Australian J. Phys.* **52**, 155

3. Goodenough, J.B. and Zhou J.-S. (1999) *Physics and Chemistry of Transition-Metal Oxides*, Proc. 29th Taniguchi Symposium, Springer Solid State Sciences **125**, H. Fukuyama and N. Nagaosa, eds. (Springer-Verlag, Berlin,) p. 9

4. Onada, M., Ohta, H, and Nagasawa, H., (1991) " Metallic properties of perovskite $SrVO_3$," *Solid State Commun.* **79**, 281

5. Nguyen H.C. and Goodenough, J.B. (1995) "Localized-Itinerant Electronic Transition in the Perovskite System $La_{1-x}Ca_xVO_3$," *Phys. Rev.* **B52**, 8776

6. Fujimori, A. *et al,* (1992) " Evolution of the spectral function in Mott-Hubbard systems with d^1 configration" *Phys. Rev. Lett* **69**, 1796

7. Inoue, I.H. *et al,* (1995) " Systematic development of the spectral function in the $3d^1$ Mott-Hubbard system $Ca_{1-x}Sr_xVO_3$" *Phys. Rev. Lett.* **74**, 2539

8. Zhou, J.-S. and Goodenough, J.B. (1996) "Heterogeneous electronic structure in $CaVO_3$," *Phys. Rev.* **B54**, 13393

9. Holstein, T. (1959) *Ann. Phys. (Paris)* **8**, 325

10. Zhou, J.-S. and Goodenough, J.B. (1998) "Phonon-Assisted Double Exchange in Perovskite Manganites," *Phys. Rev. Lett* **80**, 2665

11. Zhou, J.-S., Archibald, W.B. and Goodenough, J.B. (1998) "Pressure Dependence of Thermoelectric Power in $La_{1-x}Nd_xCuO_3$," *Phys. Rev.* **B57**, R2017

12. Demazeau, G., Parent, C., Pouchard, M and Hagenmuller, P. (1972) " Sur deux nouvelles phases oxygenees du cuivre trivalent $LaCuO_3$ et $La_2Li_{0.5}Cu_{0.5}O_4$" *Mat. Res. Bull* **7**, 913

13. Mott, N.F. (1990) *Metal-Insulator Transitions,* (Taylor & Francis, Cambridge & London)

14. Goodenough, J.B. and Raccah, P.M. (1965) "Complex vs Band Formation in Perovskite Oxides," *J. Appl. Phys.* **36**, 1031

15. Lacorre, P. *et al*, (1991) " Synthesis, crystal structure, and properties of metallic $PrNiO_3$ comparison with metallic $NdNiO_3$ and semiconducting $SmNiO_3$," *J. Solid State Chem.* **96**, 225

16. Torrance, J.B., *et al*, (1992) " Systematic study of insulator-metal transitions in perovskites $RNiO_3$ (R=Pr, Nd, Sm, Eu) due to closing of charge-transfer gap" *Phys. Rev.* **B45**, 8209

17. García-Muñoz, J.L., Lacorre, P., and Cywinski, R. (1995) *Phys. Rev.* **B51**, 1597

18. Goodenough, J.B. (1996) "Covalent Exchange vs Superexchange in two Nickel Oxides," *J. Solid State Chem.* **127**, 126

19. Medarde, M. *et al*, (1998) " Giant ^{16}O-^{18}O isotope effect on the metal-insulator transition of $RNiO_3$ perovskites (R=Rare Earth)" *Phys. Rev. Lett.* **80**, 2397

20. Alonso, J.A. *et al*, (1999) " Charge disproportionation in $RNiO_3$ perovskites: simultaneous metal-insulator and structural transition in $YNiO_3$," *Phys. Rev. Lett.* **82**, 3871

21. Goodenough, J.B. (1955) "Theory of the Role of Covalence in the Perovskite-Type Manganites $(La, M(II))MnO_3$", *Phys. Rev.* **100**, 564

NEW ADVANCED MATERIALS OF La$_{1.2}$(Sr$_{1.8-x}$Ca$_x$)Mn$_2$O$_7$ WITH COLOSSAL MAGNETORESISTANCE

RU-SHI LIU AND CHIH-HUNG SHEN
Department of Chemistry, National Taiwan University, No.1,
Roosevelt Road, Section 4, Taipei, Taiwan, R.O.C.

SHU-FEN HU
National Nano Device Laboratories, Hsinchu, Taiwan, R.O.C.

JAUYN GRACE LIN
Center for Condensed Matter Sciences,
National Taiwan University, Taipei, Taiwan, R.O.C.

AND

CHAO-YUAN HUANG
Center for Condensed Matter Sciences and Department of
Physics, National Taiwan University, Taipei, Taiwan, R.O.C.

Abstract. The effects of structural, electrical and magnetic properties with the isovalent chemical substitution of Ca^{2+} into the Sr^{2+} sites in new series of the La$_{1.2}$(Sr$_{1.8-x}$Ca$_x$)Mn$_2$O$_7$ compounds (x = 0 ~ 1.8) are investigated. The highest magnetoresistance (MR) ratio [$\rho(0)/\rho(H)$] of 208 % (H = 1.5 T) at a temperature of 102 K was observed for the x = 0.4 sample. The static strain within the compound for x = 0.4 has also been found by high resolution transmission electron microscopic techniques. The Curie temperature decreases with Ca doping from 135 K to 71 K for x = 0 to 0.8 and from 355 K to 260 K for x = 1.2 to 1.8 respectively.

1. Introduction

Since the discovery of high temperature superconductivity in perovskite copper oxides, there has been revived interest in mixed valence manganese perovskites. The ABO$_3$-type manganites R$_{n+1}$Mn$_n$O$_{3n+1}$ (R = rare earth, and n = ∞), which have an insulating paramagnetic phase at high - temperature and a metallic ferromagnetic phase at low temperature, exhibit a

R. Cloots et al. (eds.), Supermaterials, 49–56.
© *2000 Kluwer Academic Publishers. Printed in the Netherlands.*

colossal magnetoresistance (CMR) in a relatively small temperature range around the Curie temperature (T_C)[1]. A suitable substitution of A^{2+} ions for R^{3+} ions results in Mn^{3+}/Mn^{4+} mixed valency, and hence the strong coupling between the magnetic ordering and the electrical conductivity demonstrates a strong relationship between the electrical resistivity and the spin alignment which has been explained by the double-exchange mechanism [2]. The Mn^{3+}/Mn^{4+} ratio and the microstructure of Mn-O network are the two key points controlling the CMR properties.

Up to now many investigations have focused on the chemical compositions of $(R,A)_{n+1}Mn_nO_{3n+1}$ (R = rare earth, A = alkali earth, and n = 2). Bilayer-structure compounds $(R, A)_3Mn_2O_7$, in which MnO_2 bilayers and $(R, A)_2O_2$ blocking layers are stacked alternatively (as shown in the inset of Fig. 1). A detailed study of the layered manganite $La_{1.2}Sr_{1.8}Mn_2O_7$ has been performed by Moritomo et al. [3]. In this compound, they have observed a ferromagnetic metal below T_C of 130 K with ahigh magnetoresistance value and a paramagnetic insulator above T_C. A detailed study of the manganites $(La,Ca)_{n+1}Mn_nO_{3n+1}$ has been performed by Asano et al [4]. In the $La_{1.2}Ca_{1.8}Mn_2O_7$ layered compound, they have found a ferromagnetic transition at T_C = 240 K and a higher (MR) ratio $[\rho(0)/\rho(H)]$; where $\rho(0)$ is the resistivity at zero magnetic field and $\rho(H)$ is the resistivity at applied magnetic field) than those of n = 3 and n = ∞ materials.

In this paper, we demonstrate the structural, electrical and magnetic properties in a series of $La_{1.2}(Sr_{1.8-x}Ca_x)Mn_2O_7$ compounds which leads to an understanding of the crossover between two ferromagnets with T_C's around 130 K in $La_{1.2}Sr_{1.8}Mn_2O_7$ and 240 K in $La_{1.2}Ca_{1.8}Mn_2O_7$.

2. Experimental

Bulk samples of $La_{1.2}(Sr_{1.8-x}Ca_x)Mn_2O_7$ ($0 \leq x \leq 1.8$) were prepared by a standard ceramic process. Stoichiometric amounts of La_2O_3, $SrCO_3$, $CaCO_3$ and MnO_2 were mixed, ground, calcined, and sintered at 1200 1500 in air [5, 6]. X-ray powder diffraction measurements were carried out with a SCINTAG (X1) diffractometer (Cu Ka radiation, λ = 1.5406) at 40 kV and 30 mA. The program of GSAS [7] was used for the Rietveld refinement in order to obtain the information of crystal structures of $La_{1.2}(Sr_{1.8-x}Ca_x)Mn_2O_7$. Electron diffraction (ED) and high resolution transmission electron microscopy (HRTEM) were carried out using a JEOL 4000EX electron microscope operated at 400 kV. Bar-shape samples were cut from the sintered pellets to be used for the standard four-probe resistivity measurements. Magnetization data were taken from a superconducting quantum interference device (SQUID) magnetometer (Quantum Design).

3. Results and Discussion

In figure 1 we show the XRD patterns of $La_{1.2}(Sr_{1.8-x}Ca_x)Mn_2O_7$ (x = 0 and 1.8) The samples can be indexed to the $Sr_3Ti_2O_7$-type structure with tetragonal unit cell (space group: I4/mmm). Based on the XRD results, the series samples of $La_{1.2}(Sr_{1.8-x}Ca_x)Mn_2O_7$ (x = 0 ~ 1.8) are all single phase [5,6]. Here, we assume that the Ca^{2+} ions substituted into the Sr^{2+} sites follow the distribution as shown by the model [8]. In figure Fig. 2 we show the lattice parameters as a function of x in $La_{1.2}(Sr_{1.8-x}Ca_x)Mn_2O_7$. The lattice constants (a and c) decrease as the increase in the Ca content. These are simply due to a manifestation of the size between Ca^{2+} [1.18 for C.N. (coordination number) = 9] and Sr^{2+} (1.31 for C.N. = 9) [9].

Intensity

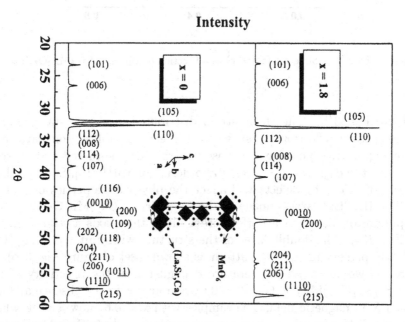

Figure 1. Powder XRD spectra of the $La_{1.2}Sr_{1.8}Mn_2O_7$ (x = 0 and 1.8) samples. The inset shows the crystal structure of $(La,Sr,Ca)_3Mn_2O_7$ projected along the c-axis. The unit cell is shown with a solid line. The MnO_6 octahedra are shaded.

The combination of electron diffraction and HRTEM allowing us to propose a structure model ($Sr_3Ti_2O_7$-type structure) which is applicable for the formula of $(R,A)_3Mn_2O_7$ (R = rare earth, A = alkali earth). In figure 3(a), we show a HRTEM lattice image taken along the [001] zone-axis direction of $La_{1.2}(Sr_{1.8-x}Ca_x)Mn_2O_7$ (x = 0.4) sample. The electron diffraction pattern is shown in the inset of figure Fig. 3(a). The lattice image along [001] of the tetragonal cell with the lattice correspondence a = b ~ a_p (a_p is the lattice constant of simple cubic perovskite structure) is shown

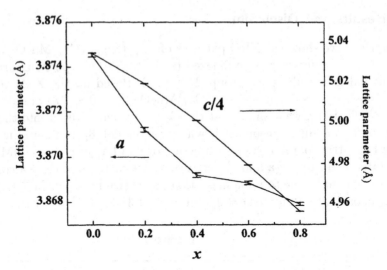

Figure 2. Lattice parameters (a and c) as a function of x in $La_{1.2}Sr_{1.8}Mn_2O_7$ (x = 0 ∼ 0.8).

in figure Fig. 3(b). The static strain distributed in the crystal produces superstructure reflection [as shown in the electron diffraction pattern of figure Fig. 3(a)] which may be caused by the charge-ordered state associated with the $d_{3x^2-r^2}/d_{3y^2-r^2}$ orbital ordering of Mn^{3+} [10]. Some of these lattice reflections or defects will affect the physical properties of materials. The HRTEM lattice image of the $La_{1.2}(Sr_{1.8-x}Ca_x)Mn_2O_7$ (x = 0.4) sample are displayed in a perfect order along the c axis as shown in figure Fig. 3(b). The double layer of the structure with unit cell along [110] is clearly present in the ED pattern as shown inset of figure Fig. 3(b). In figure 4, we show the temperature dependence of the resistivity of the $La_{1.2}(Sr_{1.8-x}Ca_x)Mn_2O_7$ (x = 0 ∼ 0.6) compounds in the absence and under a 1.5 T magnetic field. The samples with x < 0.6 show a peak which corresponds to a transition from the semiconductor-like to the metallic state on cooling. The transition temperature (T_m) at zero field decreases from 135 K for x = 0 to 102 K for x = 0.4. Moreover, the x = 0 ∼ 0.4 samples of $La_{1.2}(Sr_{1.8-x}Ca_x)Mn_2O_7$ exhibit the colossal magnetoresistance properties. The magnetoresistance(MR) = $[\rho(0)/\rho(H)]$ ratio slightly increases from ∼ 192% (135 K, 1.5 T) for x = 0 to 208% (102 K, 1.5 T) for x = 0.4. The MR ratio slightly increases from ∼ 113% (175 K, 1.5 T) for x = 1.6 to ∼ 114% (200 K, 1.5 T) for x = 1.8 is shown in the inset of figure 4. It can be seen that the MR effects of these samples exhibit complicated temperature dependence over a wide range.

In figure 5, we show the temperature dependence of the magnetiza-

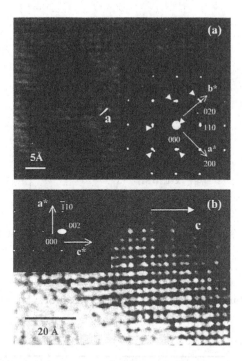

Figure 3. (a) Lattice images along the [001] direction of the $La_{1.2}(Sr_{1.4}Ca_{0.4})Mn_2O_7$ sample with diffraction magnifications. An electron diffraction pattern along the [001] direction is shown in the inset of figure 3(a). (b) HRTEM image along the c-axis for $La_{1.2}(Sr_{1.4}Ca_{0.4})Mn_2O_7$ sample. The inset of figure 3(b) shows the electron diffraction pattern along [110] of the tetragonal cell.

tion (emu/g vs. T) in a magnetic field of 0.1 T of the $La_{1.2}(Sr_{1.8-x}Ca_x)$ Mn_2O_7 (x = 0 ~ 1.8) compounds. A systematic analysis of the phase diagram of the transition temperature versus the concentration of Ca^{2+} for $La_{1.2}(Sr_{1.8-x}Ca_x)Mn_2O_7$ (x = 0 ~ 1.8) indicates that T_C decreases in different regions (x < 1.0 and x > 1.0) from 135 K to 71 K of x = 0 to 0.8 and from 355 K to 260 K of x = 1.2 to 1.8, respectively. The T_C-x curve delimits two different regions: ferromagnetic for T < T_C and paramagnetic for T > T_C as shown in figure 6. Therefore, the isovalent chemical substitution of the smaller Ca^{2+} ions into the bigger Sr^{2+} sites in the series of the $La_{1.2}(Sr_{1.8-x}Ca_x)$ Mn_2O_7 compounds (x = 0 ~ 1.8) will lead us to understand the variation of the T_C's. Detailed study on the mechanism of CMR and their correlation to the crystal structure are still underway.

4. Conclusion

The effect on structural, electrical and magnetic properties of $La_{1.2}Sr_{1.8}Mn_2$ O_7 of substituting Sr^{2+} sites with the isovalent Ca^{2+} has been investigated.

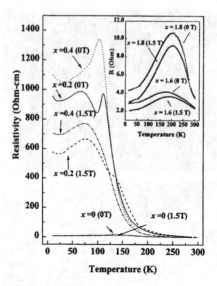

Figure 4. Temperature dependence of resistivity of $La_{1.2}Sr_{1.8}Mn_2O_7$ compounds in the absence and under a 1.5 T magnetic field.

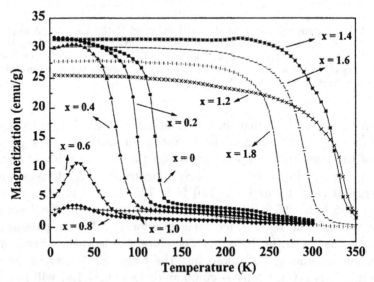

Figure 5. Temperature dependence of magnetization at 0.1 T magnetic field for the series of the $La_{1.2}Sr_{1.8}Mn_2O_7$ (x = 0 ∼ 1.8) compounds.

The highest magnetoresistance (MR) ratio $[\rho(0)/\rho(H)]$ of 208% (H = 1.5 T) at the temperature of 102 K was observed for the x = 0.4 sample. An investigation of the $La_{1.2}(Sr_{1.8-x}Ca_x)Mn_2O_7$ (x = 0 ∼ 1.8) manganites has

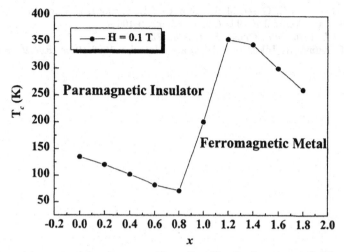

Figure 6. Magnetic phase diagram of $La_{1.2}Sr_{1.8}Mn_2O_7$ (x = 0 ~ 1.8). The circles denote the magnetic transition temperature T_C .

allowed us to establish a relationship between the ionic size and the Curie temperature.

5. Acknowledgements

The authors would like to gratefully acknowledge the assistance of the High Resolution Electron Microscopy Lab in Materials Science Center, National Tsing-hua University, supported by the National Science Council, ROC and National Tsing-hua University. This research is financially supported by the National Science Council of the Republic of China under Grant Nos NSC 88-2113-M-002-029 and NSC 88-2112-M-002-029.

References

1. Jin S, Tiefel T H, McCormack M, Fastnacht R A, Ramesh R and Chen L H 1994 *Science* **264** 413
2. Zener C 1951 *Phys. Rev.* **82** 403
3. Moritomo Y, Asamitsu A, Kuwahara H and Tokura Y 1996 *Nature* **380** 141
4. Asano H, Hayakawa J and Matsui M 1998 *Phys. Rev.* B **57** 1052
5. Liu R S, Shen C H, Lin J G, Huang C Y, Chen J M and Liu R G 1999 *J. Chem. Soc. Dalton Trans.* 923
6. Shen C H, Liu R S, Lin J G, Huang C Y, Chen J M and Liu R G 1999 *Chem.Mater.* (submitted)
7. Larson A C and Von Dreele R B 1994, Generalized Structure Analysis System, Los Alamos National Laboratory, Los Alamos, NM

8. Battle P D, Cox D E, Green M A, Millburn J E, Spring L E, Radaelli P G, Rosseinsky M J and Vente J F 1997 *Chem. Meter.* **9** 1042
9. Shannon R D 1976 *Acta. Cryst. Sect. A* **32** 751
10. Kimura T, Kumai R, Tokura Y, Li J Q and Matsui Y 1998 *Phys. Rev.B* **58** 11081

INFLUENCE OF MICROSTRUCTURE AND MAGNETIC STATE ON LOW TEMPERATURE MAGNETORESISTANCE PROPERTIES OF BULK $LA_{0.7-X}Y_XCA_{0.3}MnO_3$

F. DAMAY, J. MACMANUS-DRISCOLL,
Materials Department, Imperial College,
Prince Consort Road, London SW7 2BP, UK
L. F. COHEN
Blackett Laboratory, Imperial College,
Prince Consort Road, London SW7 2AZ, UK

The influence of the Y doping on the magnetic properties of $La_{0.7-x}Y_xCa_{0.3}MnO_3$ compounds (x = 0 ; 0.07 ; 0.1 ; 0.15) and on the low field and high field magnetoresistance responses is first investigated by means of transport, susceptibility and magnetisation measurements. It is shown that at low field the variations at 50K of the resistivity with the magnetisation are not dependent of the Y content, and that for x = 0.15 the disappearance of low field magnetoresistance is thus due to antiferromagnetic couplings leading to a low bulk magnetisation value. It is also evidenced that the high field magnetoresistance cannot be an intrinsic effect, but is related to the existence at the surface of the grain of a spin disordered layer that becomes magnetically softer with increasing yttrium content.

To investigate further the grain boundary magnetoresistance, the microstructure and physical properties (transport, magnetic, and magnetoresistance properties) of a composite ceramic consisting of a $La_{0.7}Ca_{0.3}MnO_3$ matrix to which various percentages of Y_2O_3 (in molar ratio, 1/50, 1/10, 1/5 and 1/2) have been added are studied in a second part. Even for the highest percentage of yttrium oxide, a high magnetoresistance effect of more than 50% at 170K under 0.5T is observed. Moreover, in this material the sensitivity of the resistivity to very low field (H < 0.05T) is enhanced by a factor of three compared to the pure matrix. Confirming the results obtained with $La_{0.7-x}Y_xCa_{0.3}MnO_3$, this is believed to be due to the high magnetic moment value of the composite samples at low temperature and low field.

1. Introduction

The low temperature low field magnetoresistance (LFMR), i.e., the high sensitivity of the transport properties towards magnetic field at low temperature (T ≤ 100K), encountered in polycrystalline manganese perovskites such as $La_{0.7}Ca_{0.3}MnO_3$, extensively studied for its potential applications, is attributed to the existence of the

57

R. Cloots et al. (eds.), Supermaterials, 57–65.
© 2000 *Kluwer Academic Publishers. Printed in the Netherlands.*

grain boundaries. One of the first model proposed to explain this phenomenon is based on spin-polarised tunnelling (SPT) through an insulating barrier [1]. Spin scattering at domain walls has also been considered to explain the LFMR effect [2, 3], as well as the existence at the grain boundary of a strain layer with depressed T_C [4].

Decreasing the grains size in bulks material is one of the possibility that has been found to increase the grain boundary material and thus to enhance the LFMR [5, 6]. Changing the width of the disordered grain surface layer by weakening the ferromagnetic interactions within the grain, like in $La_{0.7-x}Y_xCa_{0.3}MnO_3$, leads as well to an improved low temperature magnetoresistance [7]. However, the impact of the nature of the magnetic state in the grains surrounding the grain boundaries on the LFMR and HFMR is far from well understood : the purpose of our study is a deeper understanding of these LFMR and HFMR effects. In a first part, the magnetic properties - investigated by means of susceptibility and magnetisation measurements - of samples of the series $La_{0.7-x}Y_xCa_{0.3}MnO_3$ ($x = 0$; 0.07 ; 0.1 ; 0.15) are correlated with their low and high field low temperature magnetoresistance behaviours to determine the magnetic factors influencing the grain boundary magnetoresistance.

In a second part we tried to artificially enhance the grain boundary material and the magnetic disorder, so as to get an increased LFMR effect, by adding precipitates of Y_2O_3 ($1/10^{th}$, $1/5^{th}$, and $\frac{1}{2}$ in molar ratio) in a matrix of $La_{0.7}Ca_{0.3}MnO_3$ (LCMO).

2. Experimental

Pellets of $La_{0.7-x}Y_xCa_{0.3}MnO_3$ ($x = 0$; 0.07 ; 0.1 ; 0.15) have been prepared by standard solid state reaction. La_2O_3, Y_2O_3, CaO and MnO_2 powders in stoichiometric proportion were thoroughly mixed and heated several time at 1000°C to achieve homogeneity. An intermediate sintering step was carried on at 1300°C during 12 hrs. Pellets were then pressed under 2 t/cm^2 and sintered at 1500°C during 12 hrs. According to X-ray analysis, all samples were single phased orthorhombic perovskites. Microstructure studies reveal that the grain size decreases from ~ 8μm for the undoped sample to 2μm in average for all the Y doped samples.

The $La_{0.7}Ca_{0.3}MnO_3$ matrix used for the Y_2O_3 precipitates study was prepared by grinding and heating several time at 1000°C La_2O_3, CaO and MnO_2 powders in stoichiometric proportion. An intermediate sintering step was carried on at 1200°C during 12 hrs on pellets pressed under 5 t/cm^2. Pellets were then reground and the powder ball milled for 12hrs. Pellets were pressed under 5 t/cm^2 and sintered at 1400°C for 60 hrs. This pre-synthesised $La_{0.7}Ca_{0.3}MnO_3$ pellets were then ground and mixed thoroughly with various amounts of nanometric powder of Y_2O_3 (in molar ratio, $1/10^{th}$, $1/5^{th}$ and $\frac{1}{2}$). The resulting powder was pressed into pellets and sintered at 1400°C for 60hrs. According to the X-ray analysis, two major phases are observed in the samples : the first one is related to the orthorhombic perovskite $La_{0.7-x}Y_xCa_{0.3}MnO_3$ with lattice parameters close to those observed for $x = 0.07$. The second phase is cubic Y_2O_3 oxide ($a \sim 10.6$Å).

Magnetisation and ac-susceptibility measurements were done using a vibrating sample magnetometer or a Physical Properties Measurements System (PPMS, Quantum

Design). Transport properties were measured with a standard four point technique using silver contacts, in the earth's magnetic field or in an applied field up to 3T.

3. Results and discussion

3.1. LOW TEMPERATURE MAGNETORESISTANCE IN $La_{0.7-x}Y_xCa_{0.3}MnO_3$

Ferromagnetic interactions weaken with increasing Y content in $La_{0.7-x}Y_xCa_{0.3}MnO_3$ because of the increased distortion of the Mn-O-Mn bond and the increased disorder on the A-site of the perovskite due to the small ionic radius of Y^{3+} compared to La^{3+} [8, 9]. As a consequence, whereas the $x = 0$ sample shows a sharp transition at 250K from a paramagnetic state to a ferromagnetic state, the $x = 0.15$ sample exhibits a broad transition at a much lower temperature, $T = 95K$. The magnetisation value at 5K after zero field cooling decreases as well with increasing doping in the series [10]. All samples from $x = 0$ to $x = 0.15$ exhibit an hysteresis between the ZFC and the FC curves. Though for for $x = 0$ and $x = 0.07$ samples this hysteresis is relatively small and due to the irreversible magnetic domain walls motion, the low temperature hysteresis observed for $x = 0.15$, however, is due to an antiferromagnetic component [10].

The different magnetic states that are encountered in the series are expected to influence the low temperature magnetotransport properties of the samples. Figure 1a shows the evolution of the $\rho(H)$ curves at 50K in $La_{0.7-x}Y_xCa_{0.3}MnO_3$. At this temperature, both $x = 0$ and $x = 0.07$ samples exhibit the two slopes behaviour, both LFMR and HFMR slopes increasing with higher Y content. The slight increase ($\sim 20\%$) of the LFMR between $x = 0$ and $x = 0.07$ is likely to be due to a decrease of the grain size in the latter [5]. For $x = 0.15$ nevertheless, the LFMR effect has disappeared and the two slopes behaviour is no more observable, the resistivity decreasing smoothly with increasing magnetic field. This originates from the complete different magnetic state of the compound : the antiferromagnetic interactions present within the grains even at 3T may prevent the carrier transport across the grain boundaries during the magnetic domains rotation.

To evidence the influence of the magnetisation value on the resistivity measurements, Figure 2 shows the variation of the normalised resistivity with the bulk magnetisation, calculated from $\rho/\rho_0(H)$ and $M(H)$ loops up to 0.2T, at $T = 50K$. Figure 2 evidences the fact that at low field, the variation of the resistivity versus the global magnetisation value is roughly the same for the three compositions $x = 0$, 0.07 and 0.15, implying that the antiferromagnetic correlations are the factor limiting the magnetoresistance, because they hinder ferromagnetic ordering for $x = 0.15$.

As $x = 0$ and $x = 0.07$ have similar ferromagnetic states below their T_C (respectively of 250K and 150K), with a difference in their magnetisation value of less than 1% at 10K under 9T, we have studied $\rho(H)$ curves at a reduced temperature, $T/T_C = 0.25$ (Fig. 1b), to investigate the impact of the ferromagnetic correlations on the HFMR effect. At T/T_C, thermal spin fluctuations are similar in the samples and the strength of the ferromagnetic interactions can thus be compared.

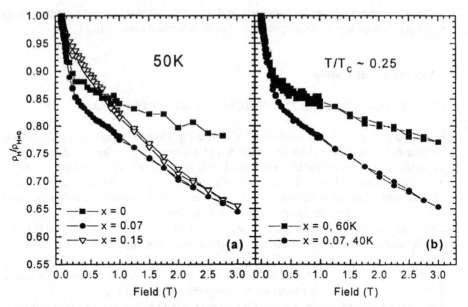

Figure 1. Normalised resistivity ρ/ρ_0 at 50K (a) and at $T/T_C \sim 0.25$ (b) versus applied magnetic field for samples of the $La_{0.7-x}Y_xCa_{0.3}MnO_3$ series ($x = 0$; 0.07 ; 0.15).

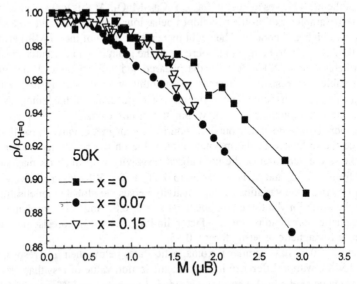

Figure 2. Normalised resistivity ρ/ρ_0 versus magnetisation (upper panel) or normalised magnetisation (lower panel) in $La_{0.7-x}Y_xCa_{0.3}MnO_3$ ($x = 0$; 0.07 ; 0.15) at 50K. Curves have been calculated from $\rho/\rho_0(H)$ and $M(H)$ loops up to 0.2T (upper panel) and from $\rho/\rho_0(H)$ and $M(H)$ loops up to 3T (lower panel).

At $T/T_C = 0.25$ (Fig. 1b), the HFMR effect is larger for $x = 0.07$ than for $x = 0$, though the global magnetisation values are similar for both samples. Parallel slopes would have been expected at the same reduced temperature - considering that the magnetic state is the same for both samples - if the HFMR effect was intrinsic to the grains. This result is consistent with the presence at the surface of the grains of a non-ferromagnetic layer as suggested by Balcells et al. [4] to explain the HFMR. The discrepancy observed between $x = 0$ and $x = 0.07$ can be related to a magnetically softer layer in the case of the latter. The actual difference of grain sizes between $x = 0$ and $x = 0.07$ is unlikely to explain the difference in their HFMR slopes, according to Balcells et al. [4].

Results indicate that yttrium doping yields to an increase of the LFMR and HFMR effects, due to a magnetically softer disordered grain boundary area for the latter. The enhanced LFMR effect observed in $x = 0.07$ compared to $x = 0$ is likely to be due to an increase of the grain boundary material in the doped compounds according to the similarity of the $\rho(M)$ curves of both compositions.

3.2. INFLUENCE OF Y_2O_3 PRECIPITATES IN $La_{0.7}Ca_{0.3}MnO_3$ ON THE MAGNETORESISTANCE PROPERTIES

3.2.1. *Microstructure*
SEM studies in combination with X-ray diffraction show that three phases can be identified : the LCMO matrix, a perovskite-type phase where yttrium has diffused (the cell parameters of this phase are close to those of $La_{0.63}Y_{0.07}Ca_{0.3}MnO_3$), and Y_2O_3 (Fig. 3).

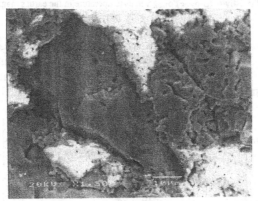

Figure 3. Microstructure of $La_{0.7}Ca_{0.3}MnO_3$ + ½ Y_2O_3 (SEM picture)

3.2.2. *Physical properties*
Resistivity versus temperature curves (Fig. 4) show, simultaneously with an increase of the resistivity at 300K with increasing yttrium content, the appearance of a second peak in addition to the peak observed at 250K that corresponds to $La_{0.7}Ca_{0.3}MnO_3$. The second peak occurs around 150K, in agreement with a $La_{0.7-x}Y_xCa_{0.3}MnO_3$ phase ($x <$ 0.1). Note that surprisingly the temperature of this second peak does not evolve linearly with the amount of Y_2O_3 in the samples, suggesting different diffusion processes.

Below 150K, there is percolation of a metallic phase, likely to be $La_{0.7-x}Y_xCa_{0.3}MnO_3$. The insulator to metal transition occurring near 150K is sharp for all the samples, which

62

suggests that the yttrium rich phase is homogeneous in composition. Microwaves results show that conduction through a metallic phase exhibiting an insulator to metal transition around 150K is real and that the second peak observed on dc resistivity curves does not result from disordered or oxygen deficient grain boundary [11].

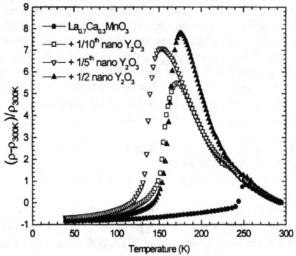

Figure 4. Normalised resistivity $(\rho-\rho_{300K})/\rho_{300K}$ versus temperature in the earth's magnetic field for all the samples (labelled on the graph) of the series.

At low field (H < 0.5T), the temperature variations of the magnetoresistance ratio (Fig. 5) for $La_{0.7}Ca_{0.3}MnO_3$ shows a peak at 250K (intrinsic magnetoresistance) followed by a regular increase of the magnetoresistance with decreasing temperature below 200K, attributed to the extrinsic magnetoresistance associated with the grain boundaries [12].

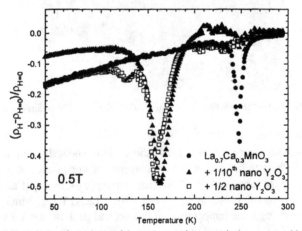

Figure 5. Temperature dependence of the magnetoresistance ratio $(\rho_{H=0.5T}-\rho_{H=0})/\rho_{H=0}$ at 0.5T for some selected samples.

The addition of precipitates increases slightly the intrinsic magnetoresistance, but affects considerably the extrinsic MR for T < 100K : for $La_{0.7}Ca_{0.3}MnO_3$ + ½ Y_2O_3, the MR ratio has a much lower value and varies more slowly with temperature than $La_{0.7}Ca_{0.3}MnO_3$. The field dependence of the magnetoresistance (Fig. 6a) shows in addition that the LFMR effect range decreases dramatically with yttrium oxide content, from [0-0.5T] for $La_{0.7}Ca_{0.3}MnO_3$ to [0-0.2T] for $La_{0.7}Ca_{0.3}MnO_3$ + ½ Y_2O_3. This narrowing of the LFMR is only observed for the highest Y_2O_3 content, but is not evidenced in $La_{0.63}Y_{0.07}Ca_{0.3}MnO_3$ (Fig. 6a). The difference of shape between the LFMR behaviours is also interesting (Fig. 6b) : $La_{0.7}Ca_{0.3}MnO_3$ shows a constant resistivity up to 20 mT, whereas for $La_{0.7}Ca_{0.3}MnO_3$ + ½ Y_2O_3 a sharp drop of resistivity is observed for H ≥ 10 mT, even though both compounds have a similar coercive field value $H_C \sim$ 30 mT. This evidences an enhanced field sensitivity in $La_{0.7}Ca_{0.3}MnO_3$ + ½ Y_2O_3 compared to $La_{0.7}Ca_{0.3}MnO_3$. The magnetisation versus temperature curves at 0.01T (Fig. 7) show as well that the magnetisation value at 40K is larger for $La_{0.7}Ca_{0.3}MnO_3$ + ½ Y_2O_3 than for $La_{0.7}Ca_{0.3}MnO_3$ or $La_{0.63}Y_{0.07}Ca_{0.3}MnO_3$: the enhanced field sensitivity at low field and low temperature seems to be related with a large magnetic moment, confirming the results obtained in Fig. 2 for $La_{0.7-x}Y_xCa_{0.3}MnO_3$.

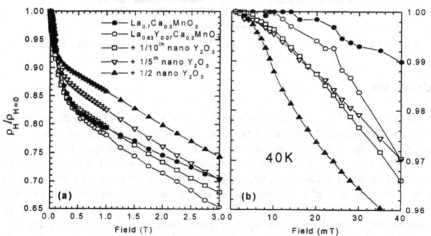

Figure 6. (a) Field dependence of the normalised resistivity at 40K for the different samples
(b) Enlargement at low field of (a)

An interesting feature of Figure 7 is that, contrary to $La_{0.55}Y_{0.15}Ca_{0.3}MnO_3$ which exhibits antiferromagnetic correlations below 70K, even the sample containing ½ in molar ratio of Y_2O_3 is ferromagnetic at low temperature. Note also that two magnetic transitions are clearly evidenced on the curve : the first one, at 250K, corresponds to $La_{0.7}Ca_{0.3}MnO_3$, the second one at 170K, broader, corresponds to $La_{0.7-x}Y_xCa_{0.3}MnO_3$. Figure 6 shows also that all the compounds containing yttrium (either by substitution or by precipitates) have same high field magnetoresistance slopes, in contrast to $La_{0.7}Ca_{0.3}MnO_3$. This result cannot be the consequence of a grain size discrepancy between the matrix and the samples containing yttrium, as such a difference would be observed only if the yttrium containing samples were nanometric [6]. This result is

64

particularly interesting considering that the high field magnetoresistance is known to be linked with a spin disordered layer at the surface of the grains. The field sensitivity of this layer would be less for $La_{0.7}Ca_{0.3}MnO_3$, but would be the same - independently of the microstructure - for all the samples containing yttrium.

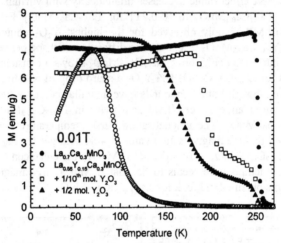

Figure 7. Magnetisation versus temperature at 0.01T for different samples of the series (ZFC curves).

4. Conclusion

In $La_{0.7-x}Y_xCa_{0.3}MnO_3$, as well in $La_{0.7}Ca_{0.3}MnO_3 + Y_2O_3$ the variations of the resistivity at low field appears to be mainly dependent of the global bulk magnetisation value resulting from the magnetic domains rotation. The enhanced sensitivity to the field of the transport properties, as well as the reduction of the low field magnetoresistance effect range observed in $La_{0.7}Ca_{0.3}MnO_3 + ½ Y_2O_3$ should be used as tools to understand the underlying physics of the phenomenon. The systematic enhancement of the high field magnetoresistance when yttrium is added to $La_{0.7}Ca_{0.3}MnO_3$ can be seen in agreement with the model proposed by Rodriguez-Martinez et al. [8], as the result of a spin disordered layer at the surface of the grain, whose contribution to the extrinsic magnetoresistance increases with magnetic disorder.

5. References

[1] Hwang, H.Y., Cheong, S.-W., Ong, N.P., and Batlogg, B. (1996), Spin-polarized intergrain tunneling in $La_{2/3}Sr_{1/3}MnO_3$, *Phys. Rev. Lett.* **77**, 2041-2044.

[2] Gupta, A., Gong, G.Q., Xiao, G., Duncombe, P.R., Lecoeur, P., Trouilloud, P., Wang, Y.Y., Dravid, V.P., and Sun, J.Z. (1996), Grain boundary effects on the magnetoresistance properties of perovskite manganite films, *Phys. Rev. B* **54**, 15629-15632.

[3] Li, X.W., Gupta, A., Xiao, G., and Gong, G.Q. (1997), Low-field magnetoresistive properties of polycrystalline and epitaxial perovskite manganite films, *Appl. Phys. Lett.* **71**, 1124-1126

[4] Balcells, Ll., Fontcuberta, J., Martínez, B., and Obradors, X. (1998), Magnetic surface effects and low temperature magnetoresistance in manganese perovskites, *J. Phys.: Condens. Matter* **10**, 1883-1890.

[5]Akther Hossain, A.K.M., Cohen, L.F., Damay, F., Berenov, A., MacManus-Driscoll, J.L., McN. Alford, N., Mathur, N.D., Blamire, M.G., and Evetts, J.E. (1999), Influence of grain size on magnetoresistance properties of bulk $La_{0.67}Ca_{0.33}MnO_{3-\delta}$, *J. Magn. Magn. Mater.* **192**, 263-270.

[6] Balcells, Ll., Fontcuberta, J., Martínez, B., and Obradors, X. (1998), High-field magnetoresistance at interfaces in manganese perovskites, *Phys. Rev. B* **58**, R14697-R14700.

[7] Fontcuberta, J., Martínez, B., Laukhin, V., Balcells, Ll., Obradors, X., Cohenca, C.H., and Jardim R. (1998), Bandwith control of the spin diffusion through interfaces and the electron-phonon coupling in magnetoresistive mangnites, *Phil. Trans. R. Soc. London, A* **356**, 1577-1592

[8] Rodriguez-Martinez, Lide M., and Attfield, J.P. (1996), Cation disorder and size effects in magnetoresistive manganese oxides perovskites, *Phys. Rev. B* **54**, R15622-R15625.

[9] Fontcuberta, J., Martinez, B., Seffars, A., Piñol, S., García-Muñoz, J.L., and Obradors, X. (1996), Colossal magnetoresistance of ferromagnetic manganites : structural tuning and mechanisms, *Phys. Rev. Lett.* **76**, 1122-1125.

[10] Damay, F., Cohen, L.F., and Driscoll, J.L., *to be published.*

[11] Yates, K., Watine, C., and Cohen, L.F., *to be published.*

[12] Evetts, J.E., Blamire, M.G., Mathur, N.D., Isaac, S.P., Teo, B.-S., Cohen, L.F., and MacManus-Driscoll, J.L. (1998), Decfect-induced spin disorder and magnetoresistance in single-crystal and polycrystal rare-earth mangnite thin films, *Phil. Trans. R. Soc. Lond. A* **356**, 1593-1615.

TAYLORING THE PROPERTIES OF LAYERED OXIDES:

Toward High-T$_c$ Nanoengineering

DAVOR PAVUNA

Department of Physics
Ecole Polytechnique Federale de Lausanne
CH - 1015 Lausanne EPFL, Switzerland

13 years ago, appeared the famous paper[1] by Bednorz and Müller, that announced a striking discovery of high-T$_c$ superconductivity in cuprates. Some 50'000 papers later, we are still trying to understand these versatile solids and control their electronic phase diagram. Nevertheless successful oxide electronic technology is already in its infancy and should have its niche in the 21st century, even if the Si and GaAs technologies remain dominant. In a direct analogy with the succesfull (Al)GaAs band-gap engineering technology I discuss the strategy to achieve similar technology with layered oxides. This is currently hindred by some intrinsic materials problems (various types of disorder) as well as some fundamental restrictions (difficult local doping and nonhomogeneous oxygen distribution). Therefore I present some of the most relevant recent experiments and discuss open questions across a rather complex electronic phase diagram. I argue that we will gradually solve all the obstacles to fully exploit the electronic phase diagram of layered cuprates (and related solids) and eventually taylor their electronic properties and novel applications within an integrated nano-engineering technology of the 21st century.

1. Introduction: Tayloring The Properties Of Layered Oxides

The main topic of this NATO workshop is the role of supermaterials. Therefore it is useful to begin with applications of oxides and compare them to prominent electronic materials, like Si or GaAs. Complete understanding of fundamentals of a given class of electronic materials often results in a successful new technology. That may require huge interdisciplinary effort of the whole generation of scientists and engineers. At present, nanoengineering of layered oxides for electronics is in its infancy, yet should have its niche, even if the Si and GaAs technologies remain dominant. However, despite remarkable experimental progress and some 50'000 published papers, high-T$_c$ oxide research is hindred by intrinsic materials complexity of these multiconstituent phases.

R. Cloots et al. (eds.), Supermaterials, 67–75.
© 2000 *Kluwer Academic Publishers. Printed in the Netherlands.*

Given high anisotropy and a short coherence length (≈1nm) layered heterostructures should ideally be controlled and fully characterized at the nanoscale level; this is not the case in general. Consequently, even fundamental mechanisms that govern the properties in the normal and in the superconducting state are not yet fully understood.

Due to the present, rather negative view of our field by the media, I begin with applications and make a rather important point. Although many of the 'leaders' of the semiconductor industry or analysts from the Wall Street consider the field of high-T_c superconductivity an investment risk, the applications are advancing successfully and may even dominate some technologies of the 21st century. Note that the successful Si-based technology has so far acumulated more than 10^7 men-years of know-how, III-V (Ga-As) photonic technology 10^6 men-years, while all superconductivity hasn't even reached the 10^5 men-years. We clearly need at least another decade of intensive R & D before giving any definite conclusions to the global media. Especially so, as there is no doubt that the in-depth understanding of the fundamentals[2] of our field (superfluidity included) will certainly be relevant to many branches of advanced science and technology in the 3rd millenium[3,8].

To elucidate the concept of high-T_c oxide 'nano-engineering' one can use a direct analogy with the bandgap engineering of (Al)GaAs lasers[4]: the wavelength is altered by varying the Al content in an optically clean, epitaxially grown Al-Ga-As nano-layered matrix. An equivalent approach to layered high-T_c (and related) oxides requires, at the very least, the following: i) Atomically flat heteroepitaxy of electronically clean constituent blocks (sub-layers) and 'clean' interfaces. Here, considerable progress hase been made[3]. ii) Precise control of the stoichiometry and local carrier doping, across the whole electronic (magnetic) and crystallographic phase diagram. This still poses formidable difficulties and remains a profound fundamental and technological challenge. iii) Reproducible control of the superconducting gap and/or insulating barrier. This goal is often hindred by incipient, growth-induced disorder, especially the distribution of oxide vacancies, that leads to localized states and/or to (nano-)phase separation. That in turn poses a challenge to our understanding of the electrodynamic response as well as correct intepretation of the microscopic origin of the so-called 'pseudogap'[5], the exact role of point defects, twins and grain boundaries, the symmetry of the gap-parameter[2] and ultimately the understanding of the puzzling difference between hole- and electron doped cuprates. In what follows I will briefly discuss anomalous electronic properties of high-T_c cuprates.

2. Anomalous Electronic Properties of High-T_c Oxides

To fully understand difficulties of nano-engineering of layered oxides one should realize that these are i) ceramic materials, usually stable only as insulators, ii) highly anisotropic solids (except for the isotropic $Ba_{1-x}K_xBiO_3$), and iii) artificial ionic metals, obtained by doping of the parent insulating compound. Their properties are complex as compared to normal metals. For example, the normal state properties of these ionic metals are anomalous and in the underdoped regime and under very high magnetic fields these superconductors seem to exhibit phase transition directly into the insulating state ('Bobinger anomaly')[7].

As shown in Figure 1 the structure of $YBa_2Cu_3O_{7-}$ can be schematically represented as a layered structure that consists of two CuO_2 planes separated by Y site. Between these bi-layers are interlayer regions which, in the case of $YBa_2Cu_3O_7$, correspond to the CuO chains (see Figure 1). In general most cuprate superconductors can be discussed in terms of the block-reservoir and the doped CuO_2 plane and they can be artificially 'constructed' by block-by-block epitaxial growth.

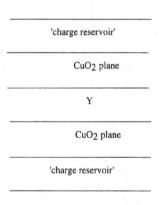

Figure 1: A model unit of layered cuprate, in this case the $YBa_2Cu_3O_{7-\delta}$.

Before discussing the anomalous electronic phase diagram I strongly emphasize that even the very best single crystals of HTSC oxides often contain various defects and imperfections like oxygen vacancies, twins, impurities ... These imperfections are not only relevant to their physical properties but possibly even essential to their basic thermodynamic stability. It may well turn out that various imperfections found in HTSC crystals are intrinsic to these materials. Note that to this day even the highest Tc phases like Hg-cuprates were not synthisized in the form of large single crystals; samples are at best <1mm in diameter.

Table 1: Most representative HTSC compounds. The index n refers to the number of CuO_2 superconducting layers within a given crystallographic structure. m refers to the number of 'chains' in the structure; m=1.5 corresponds to the case of alternating 'chains':

Compound	$T_c(K)$
$La_{2-x}M_xCuO_{4-y}$ M = Ba, Sr, Ca; x ~ 0.15, y small	38
$Nd_{2-x}Ce_xCuO_{4-y}$ (electron doped)	30
$Ba_{1-x}K_xBiO_3$ (isotropic, cubic)	30
$R_1Ba_2Cu_{2+m}O_{6+m}$ R: Y,La, Nd, Sm, Eu, Ho, Er, Tm, Lu m=1 ('123') m=1.5 ('247') m=2 ('124')	92 95 82
$Bi_2Sr_2Ca_{n-1}Cu_nO_{2n+4}$ n = 1 ('2201') n = 2 ('2212') n = 3 ('2223')	~10 85 110
$Tl_2Ba_2Ca_{n-1}Cu_nO_{2n+4}$ n = 1 ('2201') n = 2 ('2212') n = 3 ('2223')	80 100 125
$HgBa_2Ca_2Cu_3O_{10}$	133

Moreover, it is important to understand that the materials science of HTSC oxides is a non-trivial pursuit and that the understanding of phase diagrams, crystal chemistry, preparation and stability of these oxides requires an in-depth study and often hands-on experience in the laboratory. The advancement of our understanding of physics and appearance of applications depend very much on the advancements in materials research that is still very much in progress[3,6-10].

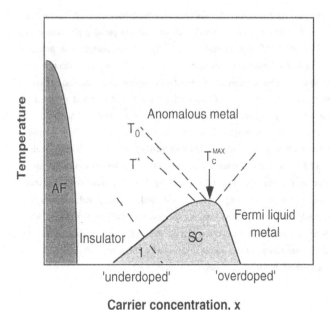

Figure 2. Schematic phase diagram of cuprate superconductors[2]. Various authors give different name or significance to various observed lines: the 'pseudogap', T^*, the 'spin-correlations' line, T_0, or the metallic Fermi-liquid region in the overdoped regime. Note anomalous behavior in region 1 where the superconductor directly transits into the insulating state in very high magnetic field[6].

Electronic Phase Diagram. True understanding of properties of high-T_C cuprates and their electronic phase diagram (Figure 2) still presents a major challenge, despite of a remarkable progress in both, sample preparation and advanced experimental techniques[2]. One of the reasons for this is that as we dope these layered oxides, we do not encounter only rather complex electronic phases; the underlying crystallographic (structural) and 'metallurgical' phase diagrams of these quaternary solids are often even more complex[3] and the disorder clearly plays an important role.

Normal State. We all seem to agree that at the left hand side we have a 2D antiferromagnetic insulator and that at the right hand side, the highly overdoped 3D perovskites tend to exhibit more Fermi-liquid-like properties. There is an agreement on the existence of the Fermi surface in the optimally doped and overdoped samples[23] (see Figure 3). Still, many details remain to be clarified, especially in the underdoped and overdoped samples. There is now also well established evidence, shown by all experiments, of the existence of two transition lines in the electronic phase diagram of

HTSC cuprates[2]: the 'pseudogap', T*, and the 'end of the spin correlations', T_0. The exact role of T_0 is discussed at length by Pines[2], while the notion of 'pseudogap' was originally proposed already in 1988 by Friedel[2], in a totally different context from presently dominant idea of a pre-formed pairs in the N-state. In a recent paper[5], Deutscher has compared gap energies, measured by different experimental techniques and has shown that these reveal the existence of two distinct energy scales: p and c. The first, determined either by angle-resolved photoemission spectroscopy or by tunnelling, is the single-particle excitation energy - the energy (per particle) required to split the paired charge-carriers that are required for superconductivity. The second energy scale is determined by Andreev reflection experiments, and Deutscher associates it with the coherence energy range of the superconducting state: the macroscopic quantum condensate of the paired charges. In the overdoped regime, p and c converge to approximately the same value, as would be the case for a BCS superconductor where pairs form and condense simultaneously. In the underdoped regime, where the pseudogap is measured, the two values diverge and p (T*) is larger than c (T_c) [5]. This indeed corresponds to the phase diagram given in Figure 2.

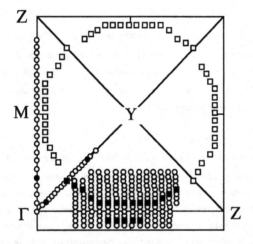

Figure 3. Experimentally determined Fermi surface : filled squares – experimental Fermi surface locations ; open squares – Fermi surface obtained by symmetry operations ; closed circles – superlattice band crossings along the Γ-M-Z and Γ-Y high symmetry directions ; open circles – locations in the Brillouin zone where the ARPES spectra were taken[23].

Metal-Insulator Transition. MI transition is not understood even in simpler solids like Si:P so it is not surprising that it remains highly disputed topic in high-T_c oxides, where anomalous behaviour was reported by Boebinger et-al[6]: deep in the underdoped regime LSCO samples directly transit from superconducting into the insulating state under very high magnetic field[5], however, this may be due to granularity of the sample. We also note that the measured band-width (obtained by ARPES) on $Sr_2CuO_2Cl_2$ – parent antiferromagnetic insulator, is consistent with the t-J model calculations, while the overall dispersion does not completely agree with this theoretical prediction[22].

Stripes. There is evidence for formation of stripes in many weakly doped perovskites[7] yet the role of stripes in the overall HTSC-scenario is still unclear: while some authors consider them vital for the mechanism, other consider them only as a stabilizing, special phase within a low-dimensional system and almost a nuisance to high-T_c.

Superconducting State. In the highly doped samples, measured properties often appear as BCS-alike[8], yet this is also highly disputed by some. In the underdoped regime the disagreement is complete and more research is needed to clarify the controversies. In the underdoped to optimally doped samples, majority of experiments indicate a dominant d-wave symmetry[2]. There seem to be, however, some notable exceptions[2,9]: there is no evidence for d-wave component in the electron doped cuprates (Maryland Center), Sharvin experiments on LSCO give a finite minimum gap (Deutscher et al) and electronic Raman experiment on Hg-2201 compound (Sacuto et al) is not compatible with d-wave but rather with extended s-wave (with nodes). Only few experiments were performed systematically in the overdoped regime[2]. Phtoemission results by Vobornik et al. in particular still remain consistent with the d-wave behavior in highly overdoped samples. Photoemission, however, cannot measure the phase of the order parameter, and therefore cannot on its own distinguish between anisotropic s and d-wave symmetry.

New Phases. Recent report[10] on possible nucleation of a 2D superconducting phase (with Tc $\approx$90K) on WO_3 single crystals surface doped with Na^+ has aparently been partly reproduced by Mihailovic et. al.[11]. The report on the apparent coexistance of superconductivity[12] (30K) and magnetism (120K) in $GdBa_2RuCu_2O_{7-y}$ seems along the line of experiments reported by Felner et. al. in 1998. We also note that striking results were reported by J.P. Locquet et. al.: they have doubled T_c (up to $\approx$50K) of LSCO films by inducing the strain into the quasi-2D matrix (by growing films on different substrates)[13]. Such strain-induced T_c-enhancement, toghether with coexistence with magnetism[12], opens numerous creative options to the artificial nanoengineering[3] of layered oxides, especially if coupled ladder-compounds eventually also exhibit superconductivity[14]. That may trace the way for future oxide technologies.

Anomalous Transport. While the main characteristics of the (anomalous) transport in cuprates seem to be well established[15] several recent results by Forro et. al.[16,17] pose a challenge to our understanding of transport. The resistivity of slightly underdoped Sr_2RuO_{4-y} superconducting (T_c=0.9K) perovskite is linear[16] over three decades of temperatures, up to 1050, yet the temperature dependence of the Hall coefficient is similar to what was measured in cuprates. This suggests that the linear temperature dependence of resistivity is not an exclusive signature of the anomalous normal state of high-T_C cuprates but rather of layered oxides in general, especially single layer perovskites, possibly independently of the magnitude of the superconducting temperature[16]. Furthermore, in single crystals of Tl-2212 [T_C =111K ($\rho_c/\rho_{ab}\approx1000$)], ρ_{ab}(T) exhibits 'usual' linear behavior and ρ_c(T) follows generally metallic-like, positive slope. However, there is a clear crossover of ρ_C(T) to semiconductor-like behavior close to T_C and, for the first time, above 500K. Under high pressures (<15 kbar) the magnitude of ρ_c strongly decreases, yet ρ_c(T) slope *does not change*[17]. That suggests pressure independent out-of-plane mechanism like in resonant tunneling in quasi-one-dimensional organic conductors, proposed by Weger[18]. Above 500K the hopping is activated hence the measured crossover in ρ_c(T)[17].

Photoemission Spectroscopy. Although advanced ARPES experiments have provided the Fermi surface of cuprates (which immediately rules out some theoretical approcahes), many profound puzzles remain. Note that recent ARPES experiments, performed on at 21 and 33 eV on well studied Bi2212 crystals by Dessau et. al., have produced somewhat different electronic features[19]; have been independently observed[20] and pose several profound questions to theorists. I also note the first report on the change of spectral signature caused by the *intentional* disorder in a Bi-2212 high-T_c cuprate by Vobornik et al.[21,22].

In conclusion, there are clearly many more puzzling results and open questions in high-T_c superconductivity, so we argue that in order to resolve numerous remaining controversies we still need many more systematic experiments on very carefully prepared and characterised samples of both, under- and over- doped, films and crystals of high-T_c oxides and related solids. And if we want to achieve the successful nano-engineering technology we also have to achieve the objectives discussed in the introduction: above all – full and reproducible control of the nano-scale stoichiometry and of local doping. Such work is currently in progress worldwide and first oxide nanoengineering company is actively pursuing some of the goals discussed here[24].

Acknowledgment I gratefully acknowledge numerous contributions of all my friends and colleagues, especially of all co-authors listed in reference 21. I also acknowledge support by the EPFL and Swiss FNRS (Bern) and thank Ivana Vobornik for Figure 3.

References

1. J. G. Bednorz and K.A. Müller, Z Phys. **B 64**, 189 (1986)

2. "The Gap Symmetry and Fluctuations in High-T_C Superconductors", Proc. NATO-ASI B371 (eds: J. Bok, G. Deutscher, D. Pavuna, S. Wolf) Kluwer-Plenum (1998)

3. D. Pavuna and I. Bozovic (eds.), *"Oxide Superconductor Physics and Nano-engineering I, II III"* no. 2058, 2697, 3481 SPIE, Bellingham (1994, 1996, 1998)

4. See "Fundamentals of Photonics" by B.E.A. Saleh, M.C. Teich, J. Wiley (1991)

5. G. Deutscher, Nature **397**, 410 (1999)

6. G. S. Boebinger et al , Phys Rev Lett **77** (27) 5417 (1996)

7. A. Bianconi (ed.), Stripes I & II, special issue of J. of Supercond. (1997 and 1999)

8. M. Cyrot and D. Pavuna, *Introduction to Superconductivity and High-T_C Materials*, World Scientific, London, Singapore, New Jersey (1992)

9. *High Temperature Superconductivity* (eds. S. Barnes et. al.) CP483 American Institute of Physics (1999); see article by R.A. Klemm et. al., p. 259

10. S. Reich and Y. Tsabba, Int. J. of Mod. Phys B **30**, in print (1999)

11. D. Mihailovic (J. Stefan, Ljubljana) private communication (1999)

12. R. Weht et. al., in ref. 9, p. 141

13. J.P. Locquet et. al. Nature **394**, 453 (1998); also J.P Locquet et. al. in ref. 3.

14. B. Normand, D.F. Agterberg, T.M. Rice, Phys. Rev. Lett. **82** (21) May 24 (1999)

15. N.P. Ong, Science **273**, 321 (1996) and references therein

16. H. Berger, L. Forró and D. Pavuna, Europhysics Letters **41** (5), 531 (1998)

17. J.P. Salvetat et. al., unpublished (1999); L. Forró, Int. J. Mod. Phys. **8**, 829 (1994)

18. M. Weger, J. Phys. Colloq. C **6**, 1456 (1978)

19. Y.D. Chuang et. al., cond-mat/9904050 preprint (1999)

20. M. Onellion (U. Wisconsin), T. Schmauder (EPFL), private communication (1999)

21. I. Vobornik, H. Berger, D. Pavuna, M. Onellion, G. Margaritondo, F. Rullier-Albenque, L. Forró, M. Grioni, Phys Rev. Lett. **82**, 3128 (1999)

22. S. La Rosa et al., Phys. Rev. B **56**, R525 (1997)

23. I. Vobornik, 'Investigation of the Electronic Properties and Correlation Effects in the Cuprates and in Related Transition Metal Oxides', D. Sci. Thesis, EPFL (1999)

24. Ivan Bozovic, Oxxel GmbH (Bremen), private communication

HOLE DOPING EFFECT ON THE MAGNETIC, ELECTRIC AND THERMOELECTRIC PROPERTIES OF La$_{2-2x}$Sr$_{1+2x}$Mn$_2$O$_7$: $(0.3 \leq x \leq 0.5)$

S. Nakamae, I. Legros, D. Colson, A. Forget and M. Ocio
Service de Physique de l'Etat Condensé, DRECAM, CEA-Saclay
91191 Gif sur Yvette, Cedex, France

J. -F. Marucco
Laboratoire des Composés Non-Stoechiométriques,
CNRS, URA 446 Bât 415, Université Paris-Sud, 91405 Orsay Cedex, France

September 1, 1999

Abstract.
We have investigated the effects of hole doping on various physical properties of double layered manganites (La$_{2-2x}$Sr$_{1+2x}$Mn$_2$O$_7$, for x = 0.3, 0.4 and 0.5). Systematic changes in the measured values of magnetization, electrical resistivity and thermoelectric power were observed as the value of x was varied. The sample with the lowest hole concentration ($x = 0.3$) appears to be most magnetic in nature and the effect of applied magnetic field decreases with increasing value of x. It has also been found that the highest hole doped sample, $x = 0.5$, possesses a magnetic structure quite different from those of $x = 0.3$ and 0.4 samples.

1. Introduction

The discovery of Colossal Magnetoresistance (CMR) in double layered manganites has spurred renewed interest in these perovskites for the fundamental physics aspects as well as the applicational possibility. Much effort has been dedicated, experimentally and theoretically, to understand the inner workings of such systems. While the double exchange mechanism clearly plays an important role in initiating a large negative magnetoresistance (MR), such process only cannot give a complete explaination to the anomoulous behavior oberved in these materials. Some recent works suggest co-existence of multiple magnetic phases as a possible origin of the observed phenomena [1], [2]. However, no consensus on the phyiscs behind these manganites has been reached.

It is a well known fact that the physical properties of CMR perovskites depend strongly on the material's hole content. Our goal in the current study is to provide additional experimental data to the on going effort in theoretical understanding of the manganites by examining the change in the physical parameters (namely, magnetization (M), resistivity (ρ) and thermoelectric power (TEP) as functions of hole content, x.

77

R. Cloots et al. (eds.), Supermaterials, 77–84.
© *2000 Kluwer Academic Publishers. Printed in the Netherlands.*

As has been discussed by many [3] [4] [5], the magnetoresistance of polycrystalline materials can be classified into two categories. The first type, here we denote as "dynamic magnetoresistance," originates from the magnetic nature intrinsic to individual grains. This type of MR appears immediately upon the application of magnetic field and saturates quickly at higher fields. The effect also disappears quickly as the temperature is lowered further away from insulator-metal transition temperature, where the thermally assisted spin alignment becomes abundant. The second type, say, "static magnetoresistance," occurs due to the presence of grain boundaries that serve as 'walls' separating magnetic domains. Such effect continues to appear in high field and low temperature range. Naturally, the separation between the two is cumbersome and often ambiguous. Thermoelectric power, on the other hand, does not suffer from such problem. As there is no electric current flowing across the grain boundaries, the TEP of individual grains are additive and thus one can observe the magnetic field effect on the intrinsic property of the materials exclusively. TEP measurements have proven practical in the study of other perovskite systems such as superconductors [6] and cubic manganites [7]. For this reason, we have included the measurements of magneto-TEP into our study.

2. Experimental

Polycrystalline $La_{2-2x}Sr_{1+2x}Mn_2O_7$ samples were used in this study in the form of powder (for magnetization measurements) and rectangular pellets (magnetoresistance and magnetothermopower studies). The hole concentration was controlled by the substitution of La^{3+} by divalent Sr cation. This increases the Mn^{4+}, and thus, the hole concentration. Three x values, 0.3 (sample 1), 0.4 (sample 2) and 0.5 (sample 3) were chosen for comparison. The x-ray diffraction and the scanning electron microscopy were used for characterization. The samples were found to be phase pure and contained randomly oriented grains.

The magnetization was measured with a commercial SQUID magnetometer. A standard four wire DC-current technique was used to measure the resistivity of the sample with the magnetic field applied perpendicular to the direction of the current flow. A heat treatement type Ag-paste was used for attaching the current and voltage leads to samples. The thermopower (normally denoted as S for 'Seebeck Coeffcient') was measured by steady state heat flow method, similar to the technique employed by Bougrine et al. [8]. Here, the magnetic field was oriented parallel to the direction of the heat flow. The obtained results were corrected for the magnetic field effect as described in [9].

3. Results and Discussion

3.1. MAGNETIZATION

The magnetization as a function of temperature of samples 1, 2 and 3 in 0.003 and 1 tesla environment (zero field cooled in both cases) are compared in Figures 1 and 2. The increase in magnetization observed at $T_N \sim 250$ K for all three samples (see Figure 1) corresponds to the transition from the paramagnetic to antiferromagnetic (AF) state. A sharp rise in the magnetization at lower temperatures, all occuring near $T_c \sim 130$ K, marks the onset of a ferromagnetic and/or a canted AF ordering of spins. The sample 3 possesses a significantly smaller mgnetization at all temperatures than those of samples 1 and 2. Furthermore, $M(T)$ of sample 3 exhibits a 'cusp' near T_N followed by a large plateau region down to T_c, in contrast to the gradual increase observed in the other two. These results reflect a change in magnetic structure of $La_{2-2x}Sr_{1+2x}Mn_2O_7$ below T_N taking place between $x = 0.4$ and 0.5.

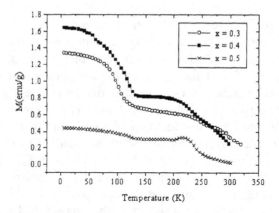

Figure 1. Magnetization vs. Temperature under $H = 0.003$ T, (field cooled)

When a stronger magnetic field (1T) is applied, T_N is no longer observable as it is overshadowed by a much larger transition at T_c. The alignment of spins appeared to be made less probable with increasing Mn^{4+} concentration as the magnitude of M climbs steadily with decreasing x (See Figure 2).

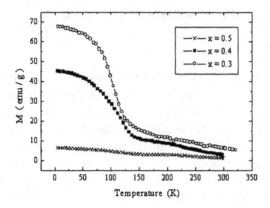

Figure 2. Magnetization vs. Temperature under $H = 1$ T (field cooled)

3.2. ELECTRICAL RESISTIVITY

In Figure 3, $\rho(T)$ measurement results in zero magnetic field of all samples are presented. The resistivity at room temperature ($T = 295$ K) is not significantly affected by their hole concentration. The transition temperature, T_c is defined here as the temperature at which $\rho(T)$ reaches its local maximum value, and is determined as 145, 125 and 132 ± 1 K for samples 1, 2 and 3, respectively. The discrepancy between the values of T_c obtained here and those found from the $M(T)$ measurements is typical of polycrystalline materials. Note that only in sample 3 (x = 0.5), $\rho(T)$ exhibits a thermal hysteresis near T_c. This is often indicative of the presence of a charge ordered (CO) state concomitant with another magnetically ordered state, as found in cubic manganite systems [2]. Our result is consistent with a recent NMR study on a series of La$_{2-2x}$Sr$_{1+2x}$Mn$_2$O$_7$ ($0.3 \leq$ x ≤ 0.5) sigle crystals by Kubota *et al.* [10], and [11], where the co-existence of a charged ordered and A-type antiferromagnetic phase was evidenced for x ≥ 0.48.

In figure 4, the magnetoresistance as a function of H of three samples measured at the respective T_c values are compared. Clearly, the sample with the lowest hole concentration ($x = 0.3$) has the largest negative dynamic MR, and the effect weakens as the value of x increases. However, as can be seen from figure 5, $\rho(T)$ of sample 1 under $H = 1$ T exhibits a field induced constant shift in resistivity even at the lowest temperature measured, indicating that the magnetoresistance stems from both static and dynamic contributions.

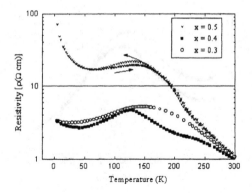

Figure 3. Resistivity vs. Temperature of all samples measured at $H = 0$ T

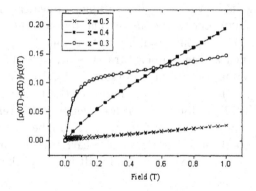

Figure 4. Magnetoresistance vs. H at $T = T_c$ of samples 1, 2 and 3

3.3. THERMOELECTRIC POWER

The results from $S(T, 0T)$ measurements are shown in Figure 6. Unlike in the case of resistivity where $\rho(280K)$ was found independent of nominal hole concentrations, $S(280 \text{ K})$ decreases, or becomes more negative with increasing x value. This general trend in the evolution of S with x is in a qualitative agreement with the results reported on $L_{1-x}A_xMnO_3$ systems [12], [13]. The value of TEP is positive for a wide range of temperatures for samples 1 and 2, whereas TEP of sample 3 remains negative for the entire range, suggesting that not only the magnetic but also the electronic structure of $x = 0.5$ system differs from that of samples with $x = 0.3$ and 0.4.

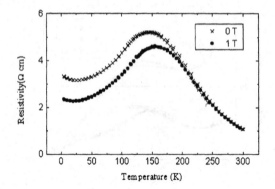

Figure 5. Resistivity vs. Temperature at $H = 0$ and 1 T of sample 2

As can be seen from the inset in Figure 6, a small thermal hysteresis was observed again in the $x = 0.5$ sample, appearing immediately below the local maximum temperature. The temperature range of the hysteresis in $S(T)$ corresponds roughly to that found in $\rho(T)$ measurement. T_{max}, defined as the temperature at which the $S(T)$ reaches its local maximum value, occurs at 155 (sample 1), 160 (sample 2) and 200 (sample 3) $\pm$ 5 K. The T_c value found in the resistivity measurements, agrees with T_{max} for $x = 0.3$; however, the difference between T_c and T_{max} widens with increasing x. This can be interpreted that a higher hole content enhances the charge carrier-spin coupling. Such "dragging effect" causes broadening and shift in the peak structures of $S(T)$.

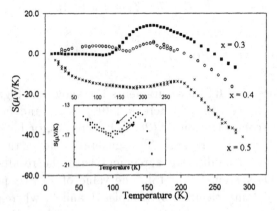

Figure 6. Seebeck Coefficient as a function of temperature of samples 1, 2 and 3 at $H = 0$ T, (Inset: the enlarged view of $S(T)$ of sample 3 around T_{max}. A small thermal hysteresis was observed.)

Due to the space limitation, here we only report the magneto-TEP data of sample 1 (See Figure 7). The more complete results and the analysis on magneto-TEP will be reported elsewhere. As expected, the magnetic field effect at low temperatures (speculated to manifestate from grain boundaries in the MR measurements) has vanished here. The field induced increase in T_{max} as well as the decrease in $S(T)$ around T_c are reminiscent of those observed in the resistivity measurement, indicating that the two phenomena are due to identical microscopic processes.

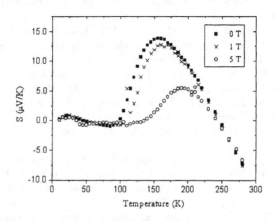

Figure 7. Seebeck Coefficient vs. Temperature at $H = 0, 1$ and 5 T of sample 2

4. Summary

The systematic effect of hole doping on the various physical parameters of $La_{2-2x}Sr_{1+2x}Mn_2O_7$ perovskites was studied. The samples with $x = 0.3$ and 0.4 demonstrate similar behavior in their magnetization, electrical resistivity and thermoelectric power, with $x = 0.4$ being more magnetic in its nature. On the other hand, the results obtained from the sample with $x = 0.5$ indicated a possibility of its unique magnetic and electric structures quite different from the other two samples. It has also be shown that the thermoelectric power is an effective tool for probing the magnetic nature of the material as it does not suffer from grain bounday effects. Currently, the measurements on TEP as a function of H is underway. The results will certainly aid us to furhter investigate the physics of charge carriers in both magnetically ordered and disordered states.

References

1. A. Moreo, S. Yunoki, and E. Dagotto. Phase separation scenario for manganese oxides and related materials. *Science*, 283: 2034–2040, 1999.

2. M. Roy, J. F. Mitchell, A. P. Ramirez and P. Schiffer. A Study of the magnetic and electrical crossover region of $La_{0.5+d}Ca_{0.5-d}MnO_3$. *Journal of Physics: Condensed Matter*, 11: 4843-4-859, 1999.

3. A. Gupta, G. Q. Gong, Gang Xiao, P. R. Duncombe, P. Lecoeur, P. Trouilloud, Y. Y. Wang, V. P. Dravid, J. Z. Sun. Grain-boundary effects on the magnetoresistance properties of perovskite manganite films. *Physical Review B*, 55(22): R15629–R15632, 1996.

4. L. I. Balcells, J. Fontcuberta, B. Martinez and X. Obradors. High-field magnetoresistance at interfaces in manganese pervoskites. *Physical Review B*, 58(22): R14697–R14700, 1998.

5. J. Klein, C. Hofener, S. Uhlenbruck, L. Alff, B. Buchner and R. Gross. On the nature of grain boundaries in the colossal magnetoresistance manganites. *Europhysics Letters*, 47(3): 371–377, 1999.

6. S. D. Obertelli, J. R. Cooper and J. L. Tallon. Systematics in the thermoelectric power of high-T_c oxides. *Physical Review B*, 46(22): 14928–14931, 1999.

7. M. Jaime, M. B. Salamon, M. Rubinstein, R. E. Treece, J. S. Horwitz, D. B. Chrisey, High-temperature thermopower in $La_{2/3}Ca_{1/3}MnO_3$ films: Evidence for polaronic transport. *Physical Review B*, 54(17): 11914–11917, 1996.

8. H. Bougrine and M. Ausloos. Highly Sensitive Method for Simultaneous Measurements of Thermal Conductivity and Thermoelectric Power: Fe and Al Examples. *Review of Scientific Instruments*, 66(1):199, 1995.

9. F. J. Blatt, A. D. Caplin, C. K. Chiang and P. A. Schroeder. Phonon drag thermopower of noble metals in a high magnetic field. *Solid State Communications*, 15:411–414, 1974.

10. M. Kubota, H. Fujioka, K. Hirota, K. Ohoyama, Y. Moritomo, H. Yoshizawa and Y. Endoh. Relation between crystal and magnetic structure of the layered manganites $La_{2-2x}Sr_{1+2x}Mn_2O_7$ (0.30 <x< 0.50). *E-print at xxx.lanl.gov*, cond-mat/9902288: 1–6, 1999.

11. M. Kubota, H. Yoshizawa, Y. Moritomo, H. Fujioka, K. Hirota and Y. Endoh. Interplay of the CE-type charge ordering and the A-type spin ordering in a half doped bilayer manganite $La_1Sr_2Mn_2O_7$. *E-print at xxx.lanl.gov*, cond-mat/9811192: 1–5, 1998.

12. A. Asamitsu, Y. Moritomo and Y. Tokura. Thermoelectric effect in $La_{1-x}Sr_xMnO_3$. *Physical Review B*, 53(6):R2952–R2955, 1996.

13. B. Fisher, L. Patlagan and G. M. Reisner. Transport properties of $L_{1-x}Sr_xMnO_3$. *Physical Review B*, 54(13):93592–9364, 1996.

LITHIUM INTERCALATION INTO THE BI-2212 PHASE THROUGH ELECTROCHEMICAL METHOD

Chemical Substitution/Doping into the Bi-Based High-Tc Superconductours

J.TAKADA

Department of Applied Chemistry, Okayama University

3-1-1 Tsushima-naka, Okayama 700-8530, JAPAN

1. Introduction

Typical subjects of my laboratory are developments of not only advanced ceramics but also metal/ ceramic composites. Our research are made on their preparations, structures and various properties such as superconducting, magnetic, electrical, mechanical properties and so on. Furthermore, we are also interested in conservation science studies of archaeological metal objects and old ceramics in Japan.

So this paper mainly treats the Bi-based high-temperature superconductors. In particular, very interesting chemical substitution and doping effects in the Bi-based high-Tc superconductors are focussed on. The effects of Pb partial substitution for Bi on the formation and the critical temperature of the 2223 phase, the annealing effects of the Pb-substituted 2223 phase, and the electrochemical Li intercalation into the 2212 phase are described. And also beautiful iron oxide red colouring of Japanese traditional potteries, which greatly influenced on the European potteries in the 17 th century, is scientifically discussed in terms of dispersion of small α-Fe_2O_3 particles.

85

R. Cloots et al. (eds.), Supermaterials, 85–99.
© *2000 Kluwer Academic Publishers. Printed in the Netherlands.*

2. Bi-based High-temperature Superconductors

2.1. BACKGROUND [1]

The Bi-Sr-Ca-Cu-O system has at least three superconducting phases expressed by their ideal compositions as "2201", "2212" and "2223". These phases have layered structures. The Tc's of the 2212 and 2223 phases are typically 80K and 105K, respectively. However, these compositions, structures, and Tc's are considerably simplified in comparison with real ones.

The typical features of the real Bi-based superconductors are as follows: (1) Their compositions are nonstoichiometric with respect to metallic elements and oxygen. (2) The atomic positions are considerably shifted from the "ideal" ones. So that each phase has one-dimensional structural modulation. (3) The Tc values of each phase depend on compositions, especially oxygen content. (4) The 2201 and 2212 phases exist as solid solutions and their monophasic samples are easily prepared [2-4]. In contrast, the 2223 phase is extremely difficult to obtain its single phase form even at present time.

The 2223 and 2212 phases are very attractive materials for practical applications. In fact, at present stage, a strong 7 Tesla magnet prepared using Ag-sheathed tapes of Pb-substituted 2223 phase was fabricated by Sumitomo Electric Industries, Ltd. [5]

However, there are still many problems to be overcome before we can reach the goal of practical use. Among these, the following points seem to be essentially important to us.

(1) To further clarify the compositional and structural features in relation to superconducting properties such as chemical substitution /doping effects.

(2) To understand the formation mechanism.

(3) To make clear other kinds of inherent features such as thermal and mechanical behaviours.

(4) To introduce strong pinning centers.

Thus, we have to continue to gather a lot of inherent information on the Pb-substituted 2223 phase and 2212 phase. In this section we discuss the Pb-substitution effects, annealing effects of the Pb-substituted 2223 phase, and electrochemical Li doping into the 2212 phase.

2.2. Pb-SUBSTITUTED 2223 PHASE

2.2.1. Pb-substitution effects

For understanding the Pb-substitution effects in the 2223 phase, the brief history of the isolation of this phase is very helpful.

On January 1988, Dr. Maeda was found to exist a new superconductors with a Tc ~105K in the Bi-Sr-Ca-Cu-O system by the electrical resistivity measurements as shown in Figure 1. [6]   Their samples with a composition of $Bi_1Sr_1Ca_1Cu_2O_z$, not the 2223 composition, however, consisted of the 2212 phase mainly so that they couldn't identify the 105K phase, that is, the 2223 phse.

Although a large number of researchers all over the world have made great efforts to obtain the single phase, usual ceramic fabrication techniques gave only mixtures of the 2212 and 2201 phases mainly. Moreover, it is noted that a small amount of the 2223 phase with poor crystallinity was formed in remarkably narrow temperature range.

We also made a careful survey in the above system to find the accurate composition of the 2223 phase and, at the same time, tried chemical substitutions for the purpose of making it "easier" to produce the 2223 phase and raising its Tc. In particular, partial substitution of Pb for Bi was studied in detail. Maybe several thousand samples were prepared and checked by X-ray diffraction analysis and electrical resistivity measurements.

On March 1988, we first found that the Pb-substitution dramatically increased the volume fraction of the 2223 phase as shown in Figure 2 [7], for a sample with a composition of Bi:Pb:Sr:Ca:Cu=1.4:0.6:2:2:3.6 heated at 845°C for a long periods of 244h. But our samples were still multiphasic, not monophasic, because small amounts of the 2212 phase and CuO were observed.

So, we have continued further detailed investigations on composition dependence for the isolation of the 2223 phase. Finally we reached two different monophasic

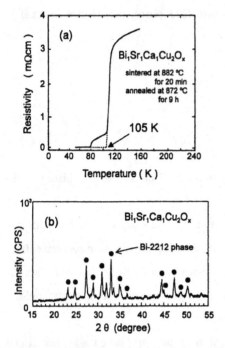

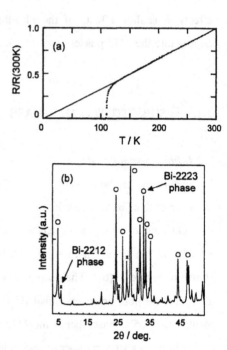

Figure 1. Electric resistivity change with temperature (a) and XRD pattern (b) in $Bi_1Sr_1Ca_1Cu_2O_x$ sample which was discovered by Maeda et al. [1] The sample mainly consisted of the 2212 phase, not the 2223 phase.

Figure 2. Electric resistivity change with temperature (a) and XRD pattern (b) in $Bi_{1.4}Pb_{0.6}Sr_2Ca_2Cu_{3.6}O_x$ sample which was found by Takada et al. [2] ◯: the 2223 phase and x: the 2212 phase. The sample mainly consisted of the 2223 phase. Partial Pb substitution dramatically promoted the formation of the 2223 phase.

compositions: $Bi_{1.8}Pb_{0.4}Sr_2Ca_2Cu_{3.2}O_z$ [8,9] and $Bi_{1.92}Pb_{0.48}Sr_2Ca_{2.2}Cu_{3.2}O_z$. [10] In addition heavily Pb substitution was found to raise *Tc* up to at least 117K. [11]

Therefore, the Pb-substitution effects in the 2223 phase are summarized as follows: remarkable promotion of the formation and its isolation, improvement of crystallinity, enlargement of the formation-temperature range ($840 \sim 860\,^\circ C$), and increase in *Tc*.

2.2.2. Annealing effects: Segregation and Dissolution Reactions

Many people believe that the Bi-based cuprate superconductors remains unchanged on annealing at low temperatures. However, it is not always true for the Pb-containing phases.

Recently, we found that Pb-substituted 2223 phase exhibited interesting and specific reactions accompanied by compositional and structural changes on heating at relatively low temperatures such as 750°C in air. [9,12,13] Furthermore, by subsequent annealing at high temperatures above 840 °C the precipitates reversibly dissolved into the mother 2223 crystals. [12,13]

Morphological confirmation of the reversible precipitation and dissolution reactions was conducted through SEM observations of a fixed region of the same sample as shown in Figure 3. Figure 3 (a) is a microstructure of the starting monophasic sample of the monophasic Pb-substituted 2223 phase with a composition $Bi_{1.8}Pb_{0.4}Sr_2Ca_2Cu_{3.2}O_Z$. In (b) one can easily see that some white precipitates indicated with arrows appeared from the mother 2223 crystals on annealing at 750 °C in air. It is quite surprising in (c) that after reheating at 850 °C in air the

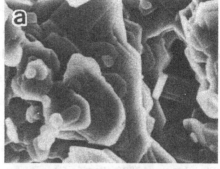
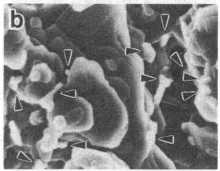
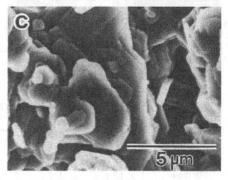

Figure 3. SEM images of a fixed region of the starting sample of the Pb-substituted 2223 phase (a), the same sample annealed at 750°C for 24h (b) and subsequently reheated at 850°C for 24h (c) in air. In (b) precipitates are shown with arrows. These particles disappeared by reheating at 850°C (c).

precipitates particles perfectly disappeared by dissolving into the mother crystals.

Similar segregation and dissolution reactions were found to be common in Pb-substituted 2212 and 2201 phases. [14-16]

Figure 4 shows the reversible annealing behaviour in XRD patterns. After annealing at 750°C, the diffraction peaks of a second phase were observed besides the peaks of the 2223 phase. The second phase was identified as $Pb_3Sr_{2.5}Bi_{0.5}Ca_2CuO_z$, which is referred to hereafter as "3321" phase. [17] It is noteworthy that the "3321" phase involves all the metallic elements containing in the Pb-substituted 2223 phase. It is interesting that the precipitation of the "3321" containing Pb^{4+} ions didn't occur in reducing atmosphere such as flowing Ar. Furthermore, the "3321" phase perfectly disappeared after subsequent heating at 850°C as shown in (c), yielding the initial Pb-substituted 2223 phase in (a).

The (119), (200) and (00$\underline{14}$) XRD peak profiles are compared in Figure 5. All these three peaks shifted towards lower angles for the sample heated at 750°C. The tetragonal lattice constants (a, c) of the pseudotetragonal 2223 phase are (5.413 Å, 37.16 Å)(a), (5.419 Å, 37.20 Å)(b), and (5.414 Å, 37.17 Å)(c). The lattice constants

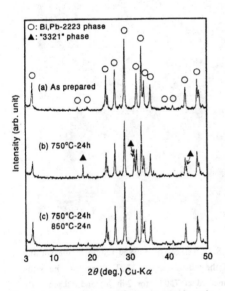

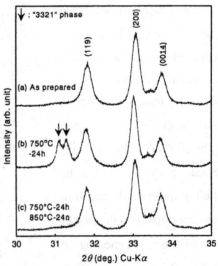

Figure 4. XRD patterns of the same sample as in Figure 3. ○: Pb-substituted 2223 phase and ▲: "3321" phase, $Pb_3Sr_{2.5}Bi_{0.5}Ca_2CuO_z$

Figure 5. (119), (200) and (00$\underline{14}$) XRD peak profiles of the same sample as in Figure 3. The peaks shown with arrows in (b) come from the "3321" phase.

increased by heating at 750°C, whereas the reheating at 850°C brought the decrease of the lattice constants up to the initial values. These changes are thus quite reversible.

As to superconducting properties the similar reversible changes were observed as shown in Figure 6. The results of susceptibility measurements are quite consistent with the results of the structural analysis. On annealing at 850 °C (c), the initial diamagnetic response was perfectly restored.

Thus it seems to be common to these Pb-substituted phases that Pb is stabilized in the superconducting phases as Pb^{2+} ions, whereas at lower temperatures in oxidizing atmosphere, it tends to be stabilized in various kinds of precipitates as Pb^{4+} ions.

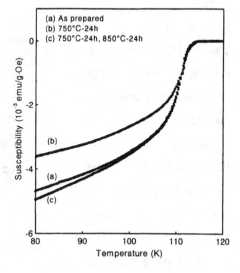

Figure 6. Temperature dependence of magnetic susceptibility of the same samples as in Figures 3-5.

2.3. ELECTROCHEMICAL LI DOPING INTO THE 2212 PHASE

Before this electrochemical doping study was started, we had studied Li doping effects on Tc and the formation process through the usual ceramics fabrication technique, that is, the solid-state reaction. [18] Then we found some interesting results: (1) The Li addition brought remarkable promotion of the formation of the 2212 phase at very low temperatures: for example the monophasic samples were obtained by heating even at 730°C in the case of 5% Li addition. This temperature is greatly lower than the formation temperatures above 830 °C in Li-free samples. (2) the maximum Li composition dissolved in the 2212 phase was only $y = 0.16$ for $Li_yBi_2Sr_{1.5}Ca_{1.5}Cu_2O_z$. (3) Tc remained almost unchanged. However, in this method a considerable Li evaporation occurred so that the control of doped-Li content was very difficult.

We recently expected that Li intercalation into the 2212 phase would occur by using an electrochemical technique. It is because the Bi-based superconductors have weekly bonded [Bi-O] double layers in their crystal structures. So it was expected that Li could be inserted into the site between the double layers through electrochemical method.

The details of our studies [19,20] are briefly described below. The monophasic disk samples of the 2212 phase with a composition $Bi_2Sr_{1.5}Ca_{1.5}Cu_3O_z$ were prepared by a usual solid-state reaction. The electrochemical reaction was performed in a galvanic cell

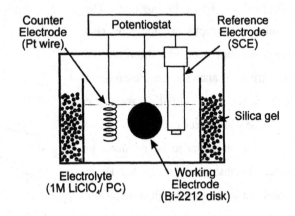

Figure 7. Schematic illustration of a galvanic cell used.

using 1.0M $LiClO_4$ dissolved in propylene carbonate as an electrolyte. The electrochemical cell for Li-doping consists of three electrodes as shown in Figure 7. A 2212 phase disk was used as the working electrode, a saturated calomel electrode as the reference electrode, and a platinum wire as the counter electrode. The samples were treated at $-1.0V$ for various periods at room temperature. The Li contents in the samples of $Li_xBi_2Sr_{1.5}Ca_{1.5}Cu_2O_z$ were determined by ICP analysis.

Figure 8 shows powder XRD patterns of non-doped sample (a) and two Li-doped samples (b) and (c) electrochemically reacted for 3.5h and 60h, respectively. The Li-doped samples keep the 2212 structure and also that any secondary phase did not form. In Figure 9, the Li-doping was found to shift the diffraction peaks towards lower angles, indicating a lattice expansion.

The change in lattice parameters with Li content is shown in Figure 10. Lattice parameters a and c increased almost linearly as the Li content increased. This result distinctly reveals that electrochemical Li-doping into the 2212 phase occurred. Furthermore, it is noteworthy that the maximum Li-content by electrochemical doping of $x = 0.50$ was markedly larger than that of 0.16 by the solid-state reaction.

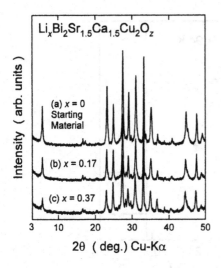

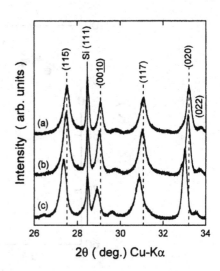

Figure 8. XRD patterns of the starting $Bi_2Sr_{1.5}Ca_{1.5}Cu_2O_z$ samples (a) and Li-doped samples electrochemically reacted for 3.5h(b) and 60h(c). All XRD peaks come from the 2212 phase.

Figure 9. Shift of the (115), (00$\underline{10}$), (117), (020), and (022) XRD peaks after the electrochemical Li-doping in the same samples as shown in Figure 8.

A dramatic change in Tc with doped Li content is clearly shown in Figure 11. An optimum Li content around x =0.12 yields Tc=92K. After passing a maximum, Tc rapidly decreases as the Li content increases. For heavy Li contents above 0.32, the doped samples exhibit non-superconducting behaviour such as insulator.

In order to discuss such remarkable Tc variation, oxygen content, average valences of Bi and Cu were determined by the coulometric titration method. As

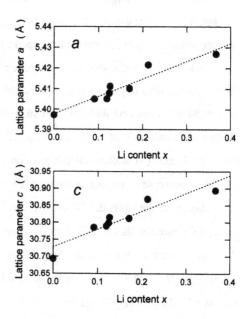

Figure 10. Variation of the lattice parameters a and c with Li content x for $Li_xBi_2Sr_{1.5}Ca_{1.5}Cu_2O_z$.

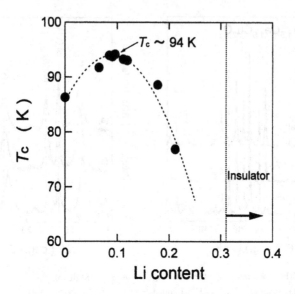

Figure 11. Change in *T*c with electrochemically doped Li content *x*.

a result, we found quite interesting change of these parameters. First, oxygen content remained constant throughout electrochemical doping. Secondly the average Bi valence decreased linearly towards 3.0. Third, the average Cu valence decreased linearly and reached approximately 3.0 around the doped Li content of $x =0.32$.

Thus, we first succeeded the electrochemical Li doping into the 2212 phase. And the following three important results were obtained: (1) extremely heavy Li-doping up to $x =0.50$ in comparison with that of 0.16 by the solid-state reaction, (2) dramatic *T*c-change: maximum *T*c of 92K around $x =0.12$ and non-superconducting properties above 0.32, and (3) unchanged oxygen content throughout electrochemical reactions and linear decrease in average valences of Bi and Cu.

Therefore, the dramatic *T*c change can be explained in terms of the change of average Cu valence, that is, hole concentration.

More recently, we also confirmed that the electrochemical reactions occurred reversibly. [20] The Li previously doped by the charged reaction can be easily removed by the discharge reaction.

Thus, two key points are emphasized : the reversible electrochemical reactions are Li-intercalation and deintercalation in the 2212 phase and have great advantages of easily controlling the Li doping level at room temperature, not at high temperatures.

3. Beautiful Iron Oxide Red Colour of Japanese Traditional Pottery of Arita-wares —Secret of Colouring of Vivid "Kakiemon red" in the 17th Century

The special red colour, yellowish red, of the Arita-ware which was made by a potter named Sakaida Kakiemon in about 1650 is called "Kakiemon Red". Kakiemon artificially created the red colour of Japanese persimmon by trial and error in about 1650. The highly advanced technique of the distinctive red colouring has been a secret of Kakiemon's family until now over 400 years. The "Kakiemon Red" colouring is due to experimental eyes of the craftsman, though he didn't know science.

It is very interesting that the red colour strongly influenced on European potteries through pieces brought back by the Dutch of the East Indian Company since approximately 1680.

Iron oxides have various applications not only in modern advanced materials, but also in ancient pigments. It is well-known that as a typical modern application γ-Fe_2O_3 is used in magnetic memory of VIDEO tapes. On the other hand, ancient mankind used α-Fe_2O_3 powder as a red colour pigment of mural painting such as in LASCAUX about 15000 years ago. It is very surprised that these ancient red colours have been still vivid.

The colouring of "Kakiemon Red" is characterized by a stacking structure of two different thin glaze layers as shown in Figure 12. The lower is a milky-white glaze

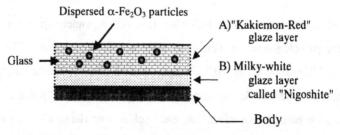

Figure 12. Schematic stacking structure of two different glaze layers on the body in the "Kakiemon Red" pottery: "Kakiemon Red" glaze layer and milky-white glaze layer. The milky-white layer is called as "Nigoshite".

layer on the body, and the upper is the "Kakiemon Red" glaze layer. But it has been unclear how the vivid colour tone of "Kakiemon Red" is made.

In 1960's Dr.Toshio Takada, who was a young chemist in Kyoto University, tried to solve the secret of the "Kakiemon red" tone, that is, yellowish red, not dark red, from the view point of materials science. [21,22] He forcused on the particle size of α-Fe$_2$O$_3$ dispersed in a glass matrix of the "Kakiemon Red" glaze layer. And then he made a lot of systematic and careful experiments using very purified and small raw materials of α-Fe$_2$O$_3$ powder. The experimental process of the "Kakiemon Red" glaze layer is schematically illustrated in Figure 13: first mixing of the raw α-Fe$_2$O$_3$ powder and a raw glass ore, next painting the mixture on the milky-white layer previously made on the body, and finally firing at low temperatures.

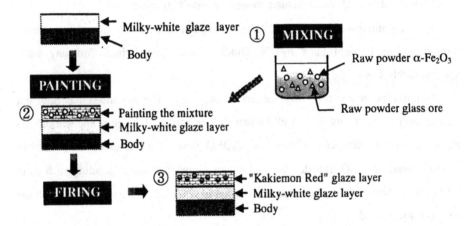

Figure 13. Experimental process of "Kakiemon Red" glaze layer : mixing of raw α-Fe$_2$O$_3$ and glass ore powders ①, painting of the mixture on the milky-white glaze layer previously made ②, and firing at 700~1000℃ ③.

As a result, it was found that the red colour tone was dependent upon mainly three factors: (A) the particle size of α-Fe$_2$O$_3$ raw material: 500Å ~2500Å, (B) the firing temperature: 600℃~1000℃, and (C) the thickness of the "Kakiemon Red" glaze layer: thin, medium, and thick. The vivid red colour corresponds to "Kakiemon Red". Series A shows the particle size effect. The glaze colour tone changes to dark red from yellowish red as the raw material particle size increased to 2500Å from 500Å. In series B, the firing temperature effect was studied: the firing at low temperatures around 700°C gave a yellowish red, while high-temperature firing resulted in dark red. Series C shows the glaze thickness effect. The yellowish red was obtained in the thin glaze. In contrast the thick glaze layer brought dark red.

In summary, the colouring of "Kakiemon Red" was found to come from the uniform dispersion of extremely small α-Fe_2O_3 particles around 500 Å in the glass matrix of thin "Kakiemon Red" glaze layer. For the dispersion of small α-Fe_2O_3 particles, the three conditions are required: to use very small raw powders, to fire at low temperatures around 700°C in order to prevent from particle growth of very small α-Fe_2O_3 raw materials around 500 Å, and to paint a thin glaze, not a thick one.

Finally, it is emphasized that on the basis of this information about iron oxide red colour , I have a research plan of conservation science to clarify the colouring of ancient mural painting (Architecture) in the old Japanese temple built in the 7 th Century.

Acknowledgements

The author thanks a great number of active colleagues of Okayama University, Kyoto University, Mie University, and the National Defence Academy for collaboration and discussion.

References

1. Takada,J., Ikeda,Y., and Takano,M. (1998) Phases and their relationships in the Bi-Pb-Sr-Ca-Cu-O system, in H.Maeda and K.Togano (eds.), *Bismuth-Based High Temperature Superconductors*, Marcel Dekker Inc., New York, pp.93-128.

2. Ikeda,Y., Ito,H., Simomura,S., Oue,Y., Inaba,K., Hiroi,Z., and Takano,M. (1989) Phases and their relations in the Bi-Sr-Cu-O system, *Physica C* **159**, 93-104.

3. Ikeda,Y., Ito,H., Simomura,S., Hiroi,Z., Takano,M., Bando,Y., Takada,J., Oda,K., Kitaguchi,H., Miura,Y., Takeda,Y., and Takada,T. (1991) Phase diagram studies of the $BiO_{1.5}$-PbO-SrO-CaO-CuO system and the formation process of the"2223 (high-Tc)" phase, *Physica C* **190**, 18- 21.

4. Grader,G.S., Gyorgy,E.M., Gallagher,P.K., O'Bryan,H.M., Johnson,D.W., Sunshine,S., Zahurak,S.M.,

Jin,S., and Sherwood,R.C. (1988) Crystallographic, thermodynamic, and transport properties of the Bi$_2$Sr$_3$-$_x$Ca$_x$Cu$_2$O$_{8+\delta}$ superconductor, *Phys. Rev. B* **38**, 757-760 .

5. Kato,T., Ohkura,K., Ueyama,M. Ohmatsu,K., Hayashi,K., and Sato,K. (1997) Development of high-Tc superconducting magnet using Ag-sheathed Bi2223 tapes, in K.Osamura and I. Hirabayashi (eds.), *Advanced. Superconducutivity* X, vol.2, Springer, Tokyo, pp.877-882.

6. Maeda,H., Tanaka,T., Fukutomi,M., and Asano,T. (1988) A new high-Tc oxide superconductor without a rare earth element, *Jpn.J.Appl.Phys.***27**, L209- L210.

7. Takano,M., Takada,J., Oda,K., Kitaguchi,H., Miura,Y., Ikeda,Y., Tomii,Y., and Mazaki,H. (1988) High-Tc phase promoted and stabilized in the Bi,Pb-Sr-Ca-Cu-O system, *Jpn.J.Appl.Phys.* **27**, L1041- L1043.

8. Ikeda,Y., Ito,H., Hiroi,Z., Takano,M., Kitaguchi,H., Takada,J., Oda,K., Miura,Y., Takeda,Y., and Takada,T. (1988) Preparation and properties of the high-Tc phase in the Pb-Bi-Sr-Ca-Cu-O system, *J.Jpn.Soc.Powder & Powder Metall.* **35**, 965-963.

9. Ikeda,Y., Takano,M., Hiroi,Z., Oda,K., Kitaguchi,H., Takada,J., Miura,Y., Takeda,Y., Yamamoto,N, and Mazaki,H. (1988) The high-Tc phase with a new modulation mode in the Bi,Pb-Sr-Ca-Cu-O system, *Jpn.J.Appl.Phys.* **27**. L2067- L2070.

10. Oda,K., Kitaguchi.,H., Takada,J., Osaka,A., Miura,Y., Ikeda,Y., Takano,M., Bando,Y., Tomii,Y., Oka,Y.,Yamamoto,N., Takeda,Y., and Mazaki.,H. (1988) Preparation of Bi-Pb-Sr-Ca-Cu-O high-Tc superconductor from coprecipitated oxalates, *J.Jpn.Soc.Powder & Powder Metall.* **35**, 959-963.

11. Takada,J., Kitaguchi, H., Egi,T., Oda,K., Miura,Y., Mazaki,H., Ikeda,Y., Hiroi,Z., Takano,M., and Tomii,Y. (1990) Superconductor with Tc=117K in the Bi-Pb-Sr-Ca-Cu-O system, *Physica C* **170**, 249-253 .

12. Kusano,Y., Nanba,T., Takada,J., Egi,T., Ikeda,Y., and Takano,M. (1994) , Segregation and dissolution reactions of the 2223 phase in the Bi,Pb-Sr-Ca-Cu-O system on annealing in air, *Physica C* **219**, 366-370.

13. Takada,J., Hosotani,K., Fujisaka,M., Kusano,Y. Nanba,T., Ikeda,Y., and Takano,M. (1992) Chemical and mechanical stability of the high-Tc phase in the Bi-Pb-Sr-Ca-Cu-O system (I)− Effects of heat treatment of crushed powder of the high-Tc phase, *J.Jpn.Soc.Powder & Powder Metall.* **39**, 779-782 .

14. Kobayashi,N., Fukuhara,M., Doi,A., Takada,T., Takada,J., and Ikeda,Y. (1994) Formation of the Pb-rich 2212 phase, *J.Jpn.Soc.Powder & Powder Metall.* **41**, 400-403.

15. Ikeda,Y., Hiroi,Z., Ito,H., Shimomura,S., Takano,M., and Bando,Y., (1989) Bi,Pb-Sr-Ca-Cu-O system including a modulation-free superconductor, *Physica C* **165**, 189-198 .

16. Niinae,Y., Kusano,Y., Nanba,T., Takada,J., Ikeda,Y., and Bando,Y. (1994) Precipitation of the Pb-compounds from Pb-sustituted Bi-2201 phase — Comparison of precipitation behavior in the 2201 and the 2223 phases, *J.Jpn.Soc.Powder & Powder Metall.* **41**, 392-395.

17. Kitaguchi,H., Takada,J., Oda,K., and Miura,Y. (1990) Equilibrium phase diagrams for the systems PbO-SrO-CuO and PbO-CaO-SrO, *J.Mater.Res.* **5**, 1397-1402.

18. Fujiwara,M., Nagae,M., Kusano,Y., Fujii,T., and Takada,J. (1996) Li doping to the 2212 phase in the Bi-Sr-Ca-Cu-O system, *Physica C* **274**, 317-322.

19. Fujiwara,M., Nagae,M., Kusano,Y., Fujii,T., Takada,J., Takeda,Y., and Ikeda,Y. (1997) Electrochemical lithium intercalation into the Bi-2212 phase, *Physica C* **279**, 219-224.

20. Nakanishi,M., Uwazumi,Y., Fujiwara,M., Takada,J., Kusano,Y., Takeda,Y., and Ikeda,Y. (1998) Lithum intercalation/deintercalation of the Bi-2212 phase by electrochemical method, *J.Jpn. Soc. Powder & Powder Metall.* **45**. 669-669.

21. Takada, T. (1958) Studies on iron red glazes, *J.Jpn.Soc.Powder & Powder Metall.* **4**, 169-186.

22. Takada, T. (1987) Studies of fine particles of iron-oxides, *J.Magn.Soc. Jpn.* **11** Supplement, xiii-xvi.

HIGH TEMPERATURE SOLUTION GROWTH OF OXIDE SINGLE CRYSTALS FOR OPTOELECTRONICS

A. MAJCHROWSKI
Institute of Applied Physics, Military University of Technology,
2 Kaliskiego Str., 00-908 Warsaw, Poland
e-mail: zielj@wat.waw.pl

Several oxide single crystals, namely CLBO, NYAB, mixed titanium sillenites, and KGW:Nd, were grown from high temperature solutions. The crystallization was carried out in multi-zone resistance furnaces in which temperature could be changed at the rates as slow as 0.01K/h due to the use of Eurotherm 906S programmers. The multi-zone construction enabled to shape the temperature gradients in a very broad range. It was shown that temperature gradients strongly influence the habit of the growing crystals. By changing the temperature gradients it was possible to decrease the number of crystallographic planes forming the interface in sillenite single crystals from nine to only one dominating plane. Growth under conditions of natural and forced convection has been observed in case of sillenite crystals. It was found that natural convection stimulated the formation of a core, enriched in dopants, in the central part of the growing crystals. The transition from the natural to forced convection did not change the shape of the interface, but removed the core and gave more uniform crystals.

1. Introduction

Enormous progress in optoelectronics has been made in recent years in connection with development of new materials, among which oxide single crystals still play very important role. They find many applications in such areas of optoelectronics as higher harmonics generators, optical parametric oscillators, diode pumped lasers, and optical data processors [1, 2, 3]. Most of these oxide single crystals, like borates, sillenites, double tungstates, and KTP family can not be grown with the use of standard techniques, such as Czochralski or Bridgman method, because of incongruent melting or high temperature irreversible phase transitions. High Temperature Solution Growth (HTSG) technique allows one to overcome these problems. By finding a proper solvent one can lower the temperature of crystallization. In this way the crystallization occurs below the temperature of phase transition, or the incongruent melting, which is connected with the formation of unwanted phases, does not occur [4]. Crystal growth from solutions is carried out under conditions of relatively low temperature gradients, comparing with growth from stoichiometric melts, what results in growth of unstrained crystals with developed facets. The disadvantages of the method are as follows:

101

R. Cloots et al. (eds.), Supermaterials, 101–106.
© 2000 *Kluwer Academic Publishers. Printed in the Netherlands.*

incorporation of solvent ions into the crystal, non-uniform doping, and inclusions of solvent or impurities.

The main goal of this work was to confirm the usability of HTSG technique in crystallization of several oxide materials for optoelectronic applications, namely pure and mixed titanium sillenites $Bi_{12}Ti_{1-x}M_xO_{20}$ (M=V, Ga, Pb, Cu) - BTMO, cesium lithium borate $CsLiB_6O_{10}$ - CLBO, yttrium aluminum borate $YAl_3(BO_3)_4$:Nd - NYAB, and potassium gadolinium tungstate $KGd(WO_4)_2$:Nd - KGW:Nd. Influence of growth conditions on the crystallization processes was investigated. Understanding of the processes occurring during crystallization is necessary to improve the quality of growing crystals. It was shown that despite moderate temperature gradients in the HTSG of oxide materials, flows in the melt influenced the segregation of components on the solid-liquid interface due to the high Prandtl number of molten oxides.

2. Crystal Growth

Growth of single crystals of the investigated materials was carried out by means of the Top Seeded Solution Growth (TSSG) technique from platinum crucibles. This method, due to use of oriented seeds, enables better control of crystallization than spontaneous crystallization. Such control is necessary when uniform crystals of good optical quality are demanded. Two- or three-zone resistance furnaces were used. High heat capacity of the furnaces and independent regulation of every heating zone allowed one to grow the crystals under stable temperature conditions. Temperature control and regulation system contained Eurotherm 906S programmers. The temperature inside the furnaces could be changed linearly at the rates as slow as 0.01°C /h. Multi-zone construction of the furnaces enabled one to shape the temperature gradients in wide range. It could be changed, depending on the kind of crystallized material, from 1°C /cm in case of borates to 30°C /cm in case of sillenites crystallization.

The choice of solvent is of great importance and strongly influences the HTSG. Such properties as high solubility for the crystal constituents, low volatility and viscosity, appreciable change of solubility with temperature and absence of elements, which are incorporated into growing crystal, were taken into account during selection of solvents. Most of used fluxes were so called "self fluxes" (e.g. $K_2W_2O_7$ in case of KGW crystallization) which did not introduce any additional elements into the solution.

The synthesis of the starting solutions was made very carefully. Oxide powders can contain a lot of water adsorbed on their surface, what could be the source of deviations of composition. Such deviations change the conditions of crystallization, which makes it difficult to establish the repeatability of the process. Water was removed from used oxides by heating them at 200°C for two hours. Components were thoroughly mixed and preheated to complete the reaction. Incomplete reaction of the components may lead to such phenomena as blue coloration of $KNbO_3$ crystals [5] or formation of compounds having high melting points. The latter case was observed during BTMO crystallization. Non-uniform mixing of components led to formation of $Bi_4Ti_3O_{12}$ crystals, instead of BTMO ones. As a result $Bi_4Ti_3O_{12}$ crystals, which were difficult to dissolve, created centres of spontaneous crystallization and made seeded crystallization difficult.

2.1. MIXED SILLENITES

Photorefractive crystals of sillenites are of great interest because of their applications in optical signal processing. The best-known sillenites, $Bi_{12}GeO_{20}$ (BGO) and $Bi_{12}SiO_{20}$ (BSO), melt congruently, contrary to $Bi_{12}TiO_{20}$. From the melt having stoichiometric composition (molar ratio of Bi_2O_3 and TiO_2 6:1) one can not grow BTO, but $Bi_4Ti_3O_{12}$. To obtain BTO and BTMO crystals, growth from solutions enriched in Bi_2O_3 (molar ratio 9:1) was carried out from ~850°C down to the eutectic point near 800°C. Crystals were rotated at 35 rpm and pulled out at 0.3mm/h. BTMO single crystals weighing up to 100g were obtained.

BTMO crystals were grown on [110] BSO seeds. As grown crystals had bottoms consisting of several (up to nine) {110} and {100} planes. Depending on the temperature gradients it was possible to change the habit of the growing crystals. Relatively small increase of the temperature gradient led to decrease of the number of planes. The best quality crystals had one dominant (110) plane, which formed more than 95% of the interface. Crystals with many planes showed a strong tendency to crack, while one-plane crystals were free from cracks.

Addition of copper to BTMO causes a photochromic effect (as-grown crystals change their colour after illumination). The effect enabled one to investigate the influence of flows in the melt on BTMO crystallization. In the initial part of the crystals was seen a core having more intensive coloration than other parts of the crystal. After reaching some diameter, striations appeared, which showed the shape of the interface. Next, when the diameter was still increased, both the core and striations disappeared, and the growing crystal was uniform with no defects in its bulk.

2.2. BORATES

Borate single crystals are sought for their good nonlinear optical properties. Generally they have high efficiency of higher harmonics generation, high threshold for laser radiation and transparency far into the ultraviolet. Borates exist in many structural types due to three-fold and four-fold coordination of boron atoms in boron-oxygen compounds [6].

CLBO [7] and NYAB [8] HTSG was investigated. CLBO melts congruently, but its very high viscosity makes crystallization very difficult. To diminish the viscosity 5 mol.% of $Cs_2O:Li_2O$ (molar ratio 1:1) self-flux was used. Both pulling and no-pulling growth were used with a rotation rate of 5 rpm. When pulled 2mm/day, CLBO crystals grew under a temperature gradient, which caused the slightly convex shape of the interface. When no pulling was used, the temperature gradient was decreased, so that

CLBO crystals grew below the surface of the melt. Such crystals were confined with many crystallographic faces. Both kinds of CLBO crystal were free from macroscopic defects, but crystals grown in the melt sometimes contained grains and were cracked. Second harmonic generation for 1.06 $\Box$ m YAG:Nd laser radiation was obtained in as grown crystals.

NYAB, which combines nonlinear optical and lasing properties, melts incongruently. $K_2Mo_3O_{10}$ was used as flux. The starting composition was 20 mol.% of NYAB in $K_2Mo_3O_{10}$ with 5% of neodymium in NYAB. Growth was carried out under conditions of low temperature gradients with no pulling in [001] direction. Growth lasted two-three weeks during which crystals were rotated at the rate of 100 rpm. As-grown crystals (1.5x1.5x1.0cm) contained some striations and inclusions in the centre but their outer parts were transparent. Strong absorption near 810nm and luminescence near 1060nm were found in samples cut off from the NYAB crystals. These properties show that NYAB may be used in diode-pumped self-doubling lasers.

2.3. POTASSIUM GADOLINIUM TUNGSTATE

Neodymium-doped potassium gadolinium tungstate is one of the best candidates for production of efficient diode-pumped microlasers. KGW:Nd reveals a high temperature irreversible phase transition at 1005°C. Owing to this, HTSG of KGW:Nd single crystals had to be carried out. $K_2W_2O_7$ was used as self-flux. Starting composition was 20 mol.% of Nd:KGW in $K_2W_2O_7$ with 3 and 8% of neodymium in Nd:KGW. Seeds oriented in [010] were used. Crystals were rotated at 50rpm, while the rate of pulling was 2mm/day. As-grown crystals had flat (010) bottom. Nd:KGW crystals without pulling were grown as well. They were confined with several crystallographic planes. These crystals showed tendency to crack. Strong absorption near 810nm and luminescence near 1060nm of obtained Nd:KGW crystals confirms their usability in diode pumped microchip lasers.

3. Discussion

Molten oxides are characterized by high Prandtl numbers, which means that heat in them is transported by means of convection. Any disorder of mass flow in the melt can therefore influence the conditions of crystallization on the crystal-melt interface. During growth of oxides by the Czochralski technique, under relatively high temperature gradients, one can observe a transition between natural convection, driven by hot melt flowing up alongside the walls of the crucible, and forced convection, driven by the rotation of the crystal in the central part of the crucible [9]. The transition between these two flows can cause such dramatic changes in heat transport that as a result crystals are partly remelted, and the interface changes its shape from convex to concave towards the melt.

HTSG is carried out under relatively small temperature gradients. One can not observe any changes of the interface shape, which all the time consists of flat faces. It may suggest that the transition from natural to forced convection does not happen. The experiment with mixed sillenites containing copper, which causes the photochromic effect, showed that the phenomenon does occur in HTSG of oxide materials. The existence of the core in the initial part of crystal confirms growth under conditions of natural convection. Melt flows towards the crystal, which cools it. As a result, the temperature gradient on the interface diminishes towards the centre of the crystal and causes incorporation of more dopant atoms in the centre than in the outer parts of the crystal - the distribution coefficient is a function of the temperature gradient between crystal and melt.

When the diameter of the growing crystal increases one can observe more disturbances. A centrifugal force, which lifts the melt in the central part of the crucible, starts to dominate, and the transition between the two kinds of flow begins. During this transition temperature conditions on the interface are unstable and cause such defects as striations or even cellular growth, when constitutional supercooling appears. After some time growth under conditions of forced convection occurs. The core vanishes because the temperature gradient on the interface is more uniform – a stream of hot melt strikes the entire interface from beneath. As a result uniform crystals without core or striations is obtained. Because the centrifugal force depends on rotation rate, one can choose the rotation, so that the transition between natural and forced convection occurs in the initial phase of the crystal growth and does not influence the quality of as grown crystal.

Temperature gradients influence also the habit of growing crystals. It was shown that sillenites, borates and tungstates grow more uniform, when one dominating plane forms the interface. In the case of many planes, cracking of crystals was observed. Different crystallographic planes have different coefficients of dopant distribution between crystal and melt. As a result crystals with many planes contain areas with different concentration of dopants. These inhomogeneities are a source of strains, which cause cracking of crystals. To obtain uniform crystals, a flat interface should be formed.

4. Conclusions

HTSG is a very useful technique that allows one to obtain many oxide single crystals that can not be grown by standard methods because of incongruent melting or phase transitions. The understanding of phenomena occurring during HTSG is of great importance and enables growth of uniform optoelectronic crystals having good optical quality. It was shown that choice of growth conditions (dominating convection, temperature gradients) can strongly influence the quality (core, striations) as well as the habit of growing oxide crystals.

Mixed sillenite, borate and tungstate single crystals having good optical properties were grown by means of the HTSG technique.

5. Acknowledgments

This paper was partly supported by the Polish State Committee on Science, project No. 7T08A 008 13 and by project PBW818/WAT.

6. References

1. Huignard, J.P., and Gunter, P., (1989) *Photorefractive Materials and their Applications II. Survey of Applications*, Huignard, J.P. and Gunter, P. Eds. Springer-Verlag, Berlin

2. Becker, P., (1998) Borate Materials in Nonlinear Optics, *Adv. Mat.* Vol.10, No.13, 979-992

3. Graf, T., and Balmer, J., (1995) Lasing properties of diode-laser-pumped Nd:KGW, *Optical Engineering*, Vol. 34, No. 8 2349-2352.

4. Elwell, D., and Scheel, H.J., (1975) *Crystal Growth from High-Temperature Solutions*, Academic Press, London.

5. Mierczyk, Z., Majchrowski, A., Zmija, J., (1995) Blue coloration of potassium niobate single crystals, *Solid State Crystals: Materials Science and Applications* J. Zmija, A. Rogalski, J. Zielinski, Eds., Proc. SPIE Vol. 2373, 98-101.

6. Chen, C., Wu, Y., and Li, R., (1990) The development of new NLO crystals in the borate series, *J. Cryst. Growth*, 99 790-798.

7. Mori, Y., Kuroda, I., Sasaki, T., and Nakai, S., (1995) Nonlinear optical properties of cesium lithium borate, *Jpn. J. Appl. Phys.*, Vol. 34, L296-L298.

8. Iwai, M., Mori, Y., Sasaki, T., Nakai, S., Sarakura, N., Liu, Z., and Segawa,Y., (1995) Growth and optical characterization of Cr:YAB and Cr:YGAB crystal for new tunable and self-frequency doubling laser, *Jpn. J. Appl. Phys.*, Vol. 34, 2338-2343.

9. Majchrowski, A., and Zmija, J., (1993) Czochralski growth of oxide single crystals under conditions of forced convection in the melt, *Materials Science and Engineering*, A173, 19-22.

COPPER SIGNALS BY EPR IN $La_{1.85}M_{0.15}CuO_4$ DILUTED IN $LaMAlO_4$ (M=Sr, Ba)

P. Odier
Laboratoire de Cristallographie-CNRS, BP 166 Grenoble cedex 09, France.

N. Raimboux and P. Simon
Centre de Recherche sur les Matériaux à Haute Température-CNRS, 45071 Orléans Cedex 2, France.

N. Bobrysheva and M. Mikhailova
St Petersbourg State University, Department of Chemistry, St Petersbourg, Petrodvorets 198904 Russia.

Abstract

$La_{1.85}M_{0.15}CuO_4$ (M = Ba, Sr) have been dissolved in $LaMAlO_4$ diamagnetic matrices and the corresponding EPR signals have been studied. When the copper is in a tetrahedral site ($LaBaAlO_4$ matrix, β-K_2SO_4), its hyperfine structure is observed (for dilutions in the range of 1% and 3%). When copper is in octahedral position of structure ($LaSrAlO_4$ matrix, K_2NiF_4), a signal of orthorhombic symmetry is seen. It is broadened by magnetic interactions that can develop in this structure and the h.f. structure is not resolved. An underlying contribution (S=3/2) assigned to clusters involving Cu^{3+} must be considered in both cases. It is not presently known whether this could be related to superconductivity or not.

1. Introduction

Identifying the copper states and the detailed short-range inter-atomic interactions in copper oxide superconductors is still un-resolved, although it should be a master piece in the puzzle of high Tc's superconductivity mechanisms. While EPR of copper ions could have been a very powerful tool in the study of local structure and spin dynamics of these compounds the fast relaxation time of copper species related to their magnetic surrounding had impeded for a long time reliable data to be obtained [1]. Only recently [2], intrinsic EPR signals of $La_{2-x}Sr_xCuO_4$ have been reported in single crystals for $x \leq 0.075$ and below 225 K. The signals are very large (6 kG at 200K) and the authors attributed the EPR absorption to magnetic centers created by p holes doped by Sr ions (or excess oxygen). In fact they did not observe the p paramagnetic center itself but a three-spin polaron built up by one O-hole spin interacting with two adjacent Cu spins. This picture assigning a spin contribution on the oxygen neighbor is confirmed by time of flight neutron polarised diffusion [3].

R. Cloots et al. (eds.), Supermaterials, 107–114.

$La_{2-x}Sr_xCuO_4$, x = 0.15 can be dissolved in $LaSrAlO_4$, an isotypic matrix (K_2NiF_4) which is also diamagnetic [4]. This is very interesting since one can then perform a magnetic dilution of the superconducting phase. Even more interesting, the solid solution shows superconductivity up to 37 K, a property preserved up to a dilution reaching 90 mol % in the diamagnetic phase [5]. From magnetic and superconducting properties, the authors suggest the presence of clusters of Cu(II)-O-Cu(III)-O-Cu(II) having a limited size of 0.78nm at the highest dilution and the same doping as in $La_{1.85}Sr_{0.15}CuO_4$.

In a similar way, the Ba doped compounds, $La_{1.85}Ba_{0.15}CuO_4$, also superconducting (37 K), can be diluted [6] in the diamagnetic matrix $LaBaAlO_4$. This matrix has a different structure β-K_2SO_4 where Al^{3+} is tetrahedrically coordinated by oxygen. The solid solution y($La_{1.85}Ba_{0.15}CuO_4$)-(1-y) $LaBaAlO_4$ is β-K_2SO_4 in the diluted range, i.e. up to y=0.30, and become K_2NiF_4 type above 0.5. In the primary field, the copper is then inserted in isolated tetrahedra. Curiously the authors observed that despite the tetrahedra position for copper, the superconductivity is preserved in the diluted range (at least down to y = 0.12) with Tc depressed to 14 K (measured by microwave absorption). They assumed the existence of elementary units similar to those found in $La_{1.85}Sr_{0.15}CuO_4$- $LaSrAlO_4$ solid solutions. These clusters must involve Cu(II)-O-Cu(III) units but in a tetrahedral environment.

In this context, it was very interesting to perform EPR studies of these solid solutions with the hope to identify a signal in the diluted part of the solution. This is the aim of this paper to report on such studies.

2. Experimental

The solid solutions y($La_{1.85}M_{0.15}CuO_4$)-(1-y) $LaMAlO_4$ (M = Ba, Sr) have been synthesized either from appropriate mixtures of La_2O_3, MCO_3 (M = Ba, Sr) CuO and Al_2O_3 that were pre-calcined at 1200 K or from decomposition of an equimolar mixture of nitrates giving a fine mixture of oxides calcined at 1200K. The resulting powders were ground, pressed into pellets and calcined in alumina crucibles at 1670 K during 30h. The temperature was slowly raised to prevent melting of copper oxides. X-ray diffraction was used to characterize the structures.

EPR spectra have been recorded on powders with a Bruker ER200D device working in X band (9.4 GHz) with a standard TE102 cavity. It is equipped with a Dewar for cooling down to ~100 K. In a limited number of cases we have used a Q band (34.3 GHz) with a cylindrical TE 011 cavity. In any case, the powders have been inserted in a pure silica tube which contribution when empty has been systematically checked to know the response of the spectrometer without the sample. It is always negligible. The microwave power was set typically to 200 mW for X band and 45 mW for Q band. Only a few mg of powder is necessary, the recording of the spectrum have been repeated several times to check their repeatability which is excellent. Four samples have been studied with y = 0.01 and 0.03 for M = Ba, and y = 0.05 and 0.5 for M = Sr.

3. Results and discussion

Figure 1. shows the spectra of the four samples recorded in X band at room temperature. While the spectra were registered from 0 to 7000 G, we only show the relevant part between 2000-4000 G where significant contributions come out. The spectra of samples containing Sr are significantly different from those containing Ba.

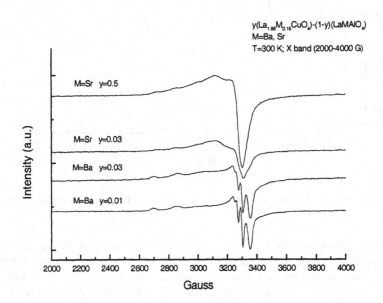

Figure 1. EPR spectrum in X band for samples of $y(La_{1.85}M_{0.15}CuO_4)-(1-y)$ LaMAlO$_4$ (M = Sr, Ba) solid solutions. Spectrum recorded at 300 K.

For Sr samples, the absorption is composed of several wide contributions that do not change significantly upon cooling, except an increases of intensity with decreasing T indicting a Curie behaviour of the paramagnetic species. The comparison between X band and Q band spectra, figure 2., clearly shows that this signal is in fact composed of two contributions better resolved in Q band. The narrow feature at 12 070 G is attributed to impurities.

110

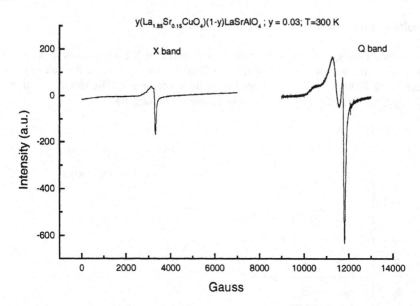

Figure 2. X band and Q band for y(La$_{1.85}$Sr$_{0.15}$CuO$_4$)-(1-y) LaSrAlO$_4$, y = 0.03, T = 300 K.

Figure 3 shows the spectra of Ba samples at two temperatures (300 K and 127 K). There is no main changes except an increasing intensity for decreasing temperatures as for Sr samples. Altogether five peaks are visible but the main signal is deformed suggesting the contributions from other peaks. At larger g (2.2-2.4) three of four contributions can be distinguished in addition. On figure 4, one compares both Ba and Sr based compounds registered in Q band. First we do not see the complex structure observed in X band for the Ba samples. Secondly, as for Sr samples, their are two main contributions, one typical of Cu^{2+} and a second one but narrower located at 11700G. It is then concluded that in both samples two contributions should be taken into account.

In figure 5, we have compared the recorded spectra to calculated one using the program WinSinfonia from Bruker. In the case of y(La$_{1.85}$Ba$_{0.15}$CuO$_4$)-(1-y) LaBaAlO$_4$ we have used the spectrum of fig.3 (y = 0.01) and data recorded at 127 K. This spectra has the typical signature of the hyperfine structure of Cu^{2+} ions and the spectrum calculated with g$_x$ = g$_y$ = 2.048 and g$_z$ = 2.29 and the hyperfine contribution (I = 3/2 and coupling constants A$_{xx}$ = A$_{yy}$ = 33 G, A$_{zz}$ = 170 G) reproduces most features of the experimental spectrum. However a second contribution must be added to better simulate the spectrum. We find that the absorption from a spin having S = 3/2, with an isotropic g factor (g$_{isotropic}$= 2.09 and a crystal field D = 400 G) summed up with the previous h.f. contribution reproduces qualitatively the spectrum if their respective weight are 0.9 for the Cu^{2+} part and 0.1 for the S = 3/2 part. Attempts to simulate the Sr species are less satisfying for the moment because of the very broad signal. Clearly, the Sr signal is due to a spin S=1/2 in an orthorhombic local symmetry to which a spin S = 3/2 contribution must be added. Even this permit to reconstruct qualitatively the shape of the Q band part.

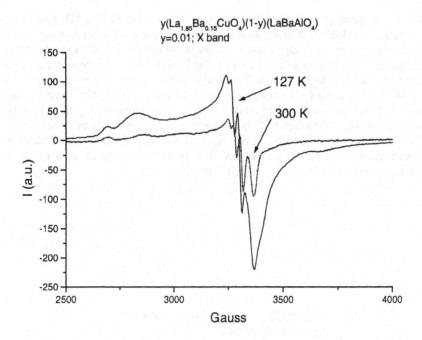

Figure 3. Effect of temperature on EPR spectrum (X band) from $y(La_{1.85}Ba_{0.15}CuO_4)$-(1-y) LaBaAlO_4, y = 0.01.

We arrive then to the conclusion that for both Ba and Sr compounds two systems of spins must be considered: one spin ½ and an another 3/2 which contribution is much smaller (~1/10). In the case of Ba compounds the main part comes from Cu^{2+}showing the h.f. structure. For Sr compounds magnetic interactions broaden the line in such a way that the hyperfine structure is not resolved. In Q band the hyperfine structure is not resolved because of dipolar interactions increasing with the magnetic field.

The complex signal of copper in the case of barium compounds must be linked with the tetrahedral symmetry occupied by copper ions in the solid solution $y(La_{1.85}Ba_{0.15}CuO_4)$-(1-y) LaBaAlO_4. While Copper ion is generally octahedrally coordinated, it is found in a tetrahedral site in some case (deformed spinelle $CuCr_2O_4$ for example) or in Milarite-type structure $A_xM_3M'Si_{12}O_{30}$ (A = Na, K, Rb), M = Mg, M' = Cu [7] for which EPR copper signals have been studied [8]. One note also that the h.f. structure of Cu^{2+} in Milarite was lost in high field because of dipolar interaction. The copper signals of Sr based compounds are typical of Cu^{2+} species in octahedral symmetry. The broad signal impede the hyperfine structure to be observed, it traduces local magnetic interactions. The strong difference between Ba and Sr compounds originates in their structural environment. In the K_2NiF_4 structure, long range magnetic

112

interactions can develop in CuO_2 planes. This is not possible in β-K_2SO_4 type solid solution because the Cu^{2+} in their tetrahedral are isolated from each other and very small possibilities of magnetic exchange exist. However in both cases spin 3/2 is involved, but in a small proportion. It might be linked with Cu^{3+} species that have been detected by chemical titration [9] to amount 10-20% of total copper. Cu^{3+} alone cannot support a S =3/2, then it is assumed that some clusters involving Cu^{2+}-O-Cu^{3+} species are responsible of this additional contribution. This view is supported by the magnetic properties of the solid solution [5,6]. It is presently not known whether this contribution is really associated with superconductivity reported to remain in less diluted samples. Other experiments (also at low T) are needed to be performed on samples in which traces of superconductivity would have been detected.

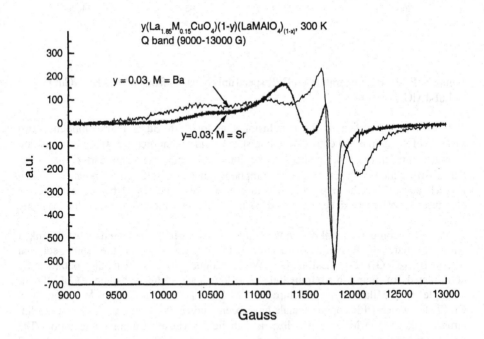

Figure 4. Comparison of barium and strontium based samples in Q band for y = 0.03.

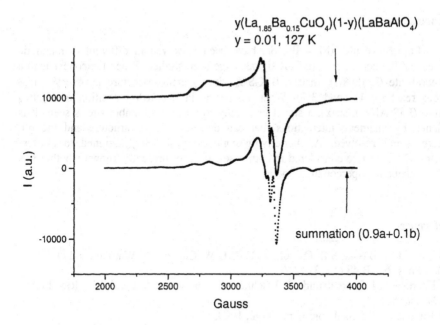

Figure 5. Comparison of calculated spectra and of a spectrum showing the h.f. structure ($y(La_{1.85}Ba_{0.15}CuO_4)$-(1-y) $LaBaAlO_4$, y = 0.01, recorded at 127 K. On top: the recorded spectra compared with the summation (0.9a+0.1b). At bottom both a and b calculated spectra.

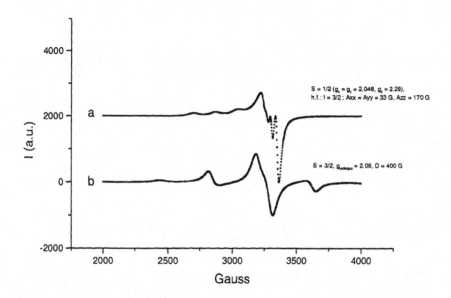

4. Conclusions

$La_{1.85}M_{0.15}CuO_4$ (M = Ba, Sr) have been dissolved in different diamagnetic matrices and the corresponding EPR signals have been studied. When the copper is in a tetrahedral site ($LaBaAlO_4$ matrix, $\beta-K_2SO_4$), its hyperfine structure is observed (for dilutions reaching 1% and 3%). When copper is in octahedral position of K_2NiF_4 structure ($LaSrAlO_4$ matrix), a signal belonging to the orthorhombic site is seen. It is broadened by magnetic interactions that can develop in this structure and the h.f. structure is not resolved. An underlying contribution (S=3/2) assigned to clusters involving Cu^{3+} must be considered in both cases. It is not presently known whether this could be related to superconductivity or not.

5. References

[1] P. Simon , J.M. Bassat, S.B. Oseroff, Z. Fisk, Q.W. Cheong, A. Wattiaux and Q. Schultz, Phys.Rev.B **48** (1993) 4216.

[2] B.I. Kochelaev, J. Sicheldsmidt, B. Elschner, W. Lemor and A. Loidl, Phy. Rev. Letter **79** (1997) 4274.

[3] L.P. Regnault, personnal communication., 1999.

[4] N. Bobrysheva, P. Mouron, J. Choisnet and N. Chezhina J. Mat.Sci.Lett. **12** (1993) 1125.

[5] N. Bobrysheva and A.J. Veinger, Zh. Obsheh. Khim. **65** (1995) 8.

[6] N. Bobrysheva, M. Mikhailova and A. Veinger, to be published.

[7] N. Nguyen, J. Choisnet and B. Raveau, J. Solid State Chem. 34 (1980) 1.

[8] J. Choisnet, D. Cornet, J.F. Hemidy, N. Guyen and Y. Dat, J. Solid State Chem. 40 (1981) 161.

[9] N. Bobrysheva, I. A. Zvereva, Y. S. Kamencev and N.V. Chezhina, USSR Inventor's Certificate ¼ 1647391 A Byul.Izobr.1991 ¼ 17.

MAGNETIZATION IN HIGH MAGNETIC FIELD AND CRITICAL MAGNETIC BEHAVIOUR IN NANOSTRUCTURED MAGNETIC OXIDES

I. NEDKOV*
Institute of Electronics
Bulgarian Academy of Sciences,
72 Tzarigradsko Chaussee,
1784 Sofia, Bulgaria

1. Introduction

The unique magnetic properties of small oxide particles have generated increasing interest due to their theoretical contributions to the study of magnetism, as well as their applications as new magnetic and magneto-optical high density recording media [1], as magnetic fluids [2], as in the treatment of cancer [3], etc.

A literature survey has produced a spectacular number of models of magnetization at the nanometric scale, although essentially concerning metals, which are different from magnetic oxides in their microstructur e. In certain ionic compounds, as the oxides of the transition group and the rare-earths, the coupled sublattices are not equivalent and the so-called superexchange interactions between the magnetic moments of different crystallographic sites are strongly antiferromagnetic (ferrimagnetic behavior). Some interesting phenomena recently observed - the reduction of the saturation magnetization and the high-field magnetization irreversibility in ferrimagnetic nanoparticles [4, 5], have again caused an increased interest in the magnetic oxides. However, the mechanism of micromagnetism in high magnetic fields in the magnetic oxide nanoparticles remains unclear. The most obvious fact is that the magnetic behavior in nanosized state is a result of contributions from both the magnetic anisotropy energy, which keeps a particle magnetized in a particular direction, and the size effects related to the spin canting. The finite-size effects in ferrioxide nanoparticles have been intensively studied to explain this behavior [6, 7]. These effects have been modeled under the assumption that the ground state is strongly influenced

R. Cloots et al. (eds.), Supermaterials, 115–127.

by microstructural defects on the surface layer with uncompensated spin magnetic moments due the exchange anisotropy behavior, and the freezing of moments into a spin-glass-like secondary magnetic phase; the thermally activated relaxation of the magnetic moments at temperatures below 50 K is responsible for this irreversibility.

On the other hand, the nano-particle can be considered as a state of matter on the borderline between the atomic and crystalline states; it possesses the encoded characteristics of the initial material in a pure form, without the defects typical for the micro-sizes in the case of complex crystal structures. A fundamental question is what kind of interactions in finite size systems are influenced by the high magnetic field. The answer requires the investigation of magnetic oxides with similar chemical composition but with anisotropy deviations such as that occuring at the structural transition due to cooperative Jahn-Teller (JT) distortion.

There are two principal ways to obtain oxide nano-particles: to diminish the size of the powder obtained by usual ceramic technology, or to synthesize directly nanosized powder. The particularities of the ball milling ferrite powders method most frequently used in practice probably cause the defects in the outer skin of the particle. The development of technologies whereby the nano-particle is directly synthesized opens possibilities to obtain a nearly *ideal* nano-particle. One could thus speak of a nanocrystal because of the similarity with the crystal growing process.

In this work, we attempt to interpret some results obtained during the investigation in high-magnetic field of $CuFe_2O_4$ nanoparticles synthesized by co-precipitation from aqueous solutions at low temperature. Two types of particles were explored: without and with cooperative JT distortion. High-field magnetization measurements ($0 - 140\ kOe$) of both types of samples were performed. Low-temperature measurements at 4.2 K demonstrated the existence of open hysteresis loops. For the sample of $CuFe_2O_4$ with JT distortion and very fine particles (grain-size 6.5 nm) noticeable hysteresis losses at room temperature were observed.

2. Experimental

The wet-chemistry capability of synthesizing nano-particulate of iron oxides (Fe_3O_4 and $\gamma - Fe_2O_3$) was demonstrated by Elisabeth Tronc and co-workers some years ago [8, 9]. The composition of the ferrite powders depends strongly on the Fe^{2+}/Fe^{3+} ratio; the co-precipitation part of the process permits an easier control of this ratio thus enabling one to use the method for obtaining a large number of different spinel structures. In the experiments reported here, the precipitate was formed by adding $NaOH$ to water solutions of $FeCl_{2.4}.H_2O$ and $CuCl_2$ mixed in strictly fixed concentration ratios (to achieve a variation of the composition with respect to the $Cu_xFe_{3-x}O_4$ initial formula), and co-precipitated in an alkaline medium at $pH > 10$. The initial solution thus obtained can be considered as a magnetic fluid of single-domain ferrite particles dispersed in a carrier liquid. Varying the pH factor of the medium where the co-precipitation process takes place can enable one to control the grain size. The precipitate was washed with deionized water to reach $pH = 7$ and then dried at a temperature of approximately $323\ K$. The technological conditions were developed to apply the method to fabricate $CuFe_2O_4$ nano-powders with and without cooperative JT distortion (this was achieved by varying the cations ratio in the initial solution), and to avoid the additional annealing of the samples. The technique allowed us to produce powders with particle size in the range of 2 to 20 nm.

X-ray phase analysis was performed by using a TUR-M62 apparatus with $CoK\alpha$ radiation. The exact position of the lines and their width were determined by computer software for deconvolution and profile analysis of diffraction patterns. The XRD data exhibited consistently a single-phase spinel structure for both types of samples. The TEM bright-field images and the grain-size statistics of the powders obtained were also analyzed.

The Moessbauer spectra were taken with an electromechanical spectrometer working at a constant acceleration mode at room temperature. A 70 mCi $^{57}Co(Cr)$ source and an $\alpha - Fe$ standard were used. The experimentally obtained spectra were computer fitted to a series of Lorentzian lines by the least-squares method [10]. The high-field magnetization measurements (0 - 140 kOe) were carried out using a water-cooled Bitter magnet with a vibrating sample magnetometer (VSM) at 4.2 K and at room temperature in the Laboratory of High Magnetic Fields and Low Temperatures in Wroclaw, Poland. The magnetic properties of the powders were investigated by means of a VSM at room temperature in fields of up to 35 kOe.

3. Results and Discussion

3.1. RESULTS

The spinel ferrites with nominal formula $Me_xFe3 - xO_4$, where $Me = Cu$, can be described as a cubic close-packed arrangement of oxygen ions with two different crystallographic sites for the cations and structure presentation as AB_2O_4 (tetrahedral-A and octahedral-B sites). The crystal structure of the particles prepared following the approaches described above at a Cu/Fe ratio 1/2 is tetragonal, while that produced at $Cu/Fe=1.5/2$ is cubic. The resulting local symmetries of the two sites are different. SEM investigation of the final powders shows the ratio $Cu/Fe=1/2$. The Cu^{2+}-cations can occupy both sites; this is a well-known Jahn-Teller ion. The

JT cooperative effect distorts the octahedral state by stretching the axial copper-oxygen bonds. The atoms form a nearly two-dimensional square grid. These layers play a central role in the $CuFe_2O_4$ spinel ferrite - it is well known that they are charge carriers responsible for the electrical properties of bulk materials [11].

The X-ray data (see fig.1a and b) show that both types of powders are single-phase, while the first one is a cubic $CuFe_2O_4$ phase. For the powder with $Fe/Cu=1/2$, the Bragg peak at $\theta \approx 27°$ indicates a tetragonal structure. According to a number of studies, Cu-spinel synthesized with the stoichiometry being strictly maintained has a tetragonal structure with lattice parameters $a = 0.820(0.824)$ nm, $c = 0.860(0.868)$ nm and $c/a = 1.05$ calculated in the (211) plane, where $d_{211}=0.2497 \pm 0.0003$. The tetragonal structure arises due to the *JT* distortion.

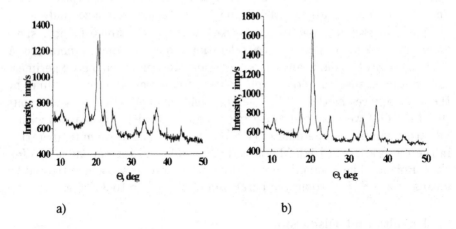

Figure 1. XRD data (CoK_α) of Cu-ferrite powders (a) sample sI with Jahn-Teller distortion and (b) sample sII without *JT*

Depending on the Cu content in the initial solution and the temperature regime, the c/a ratio can decrease and reach unity, i.e., the tetragonal structure undergoes a transformation to a cubic one. It is widely held that this transition has to do with the changes in the oxygen content in the lattice and the appearance of noticeable amounts of Cu^{1+}. The average particle size (PS) of the powders were calculated from the broadening of the peak at $\theta \approx 10.5°$ using Scherrer equation. The TEM statistics confirmed these results. Two kinds of samples were chosen for magnetization investigations; sample I (sI) with JT distortion and PS $6.5nm$, and sample II (sII) without JT distortion and PS $10nm$.

Figure 2a is a TEM micrograph of sample sI; fig.2b shows the electron diffraction (ED) pattern obtained. It can be observed that the $CuFe_2O_4$ nano-particles are very small and uniform in size (PS 6.5 nm). Agglomeration into isotropic complex of particles is also observed. The ED shows the absence of multiple-phase formation in the agglomerates; the lattice spacing in the separate grains is clearly resolved.

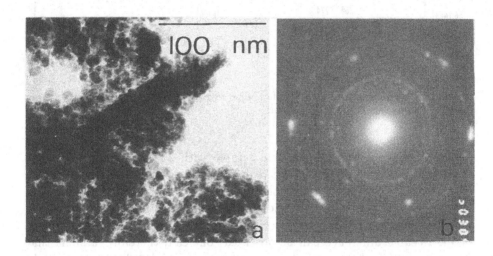

Figure 2. (a) TEM micrograph of Cu-ferrite nanocrystalline powder - sample-sI and (b) ED pattern of the aggregate formations in the powder.

The Moessbauer spectra at $295K$ of both types of samples are presented in Fig.3. The data confirm the superparamagnetic behavior of the powders. The appearance of broad lines (relaxed spectra) with a paramagnetic central doublet at channels $260 - 274$ (Fig.3a) to a level 100% of the total spectral area in the case of 6.5 nm PS of Cu-ferrite powders with JT effect is indicative of the paramagnetic nature of the material. Fig.3b displays Moessbauer spectra of the powder without JT distortion and PS about 10 nm where the superparamagnetic phase co-exists with traces (about 10%) of a ferrimagnetic phase.

The magnetic properties at magnetic fields up to 35 kOe at 295 K of the initial powders were investigated by means of a VSM. Both samples exhibited (inset in Fig.4) the so-called *without hysteresis* behavior at room temperature characteristic for the superparamagnetic state of matter.

Fig.4 presents only the first quadrant of the $M(H)$ curves at high-field (0 to 140 kOe) of both type of samples at room temperature. The

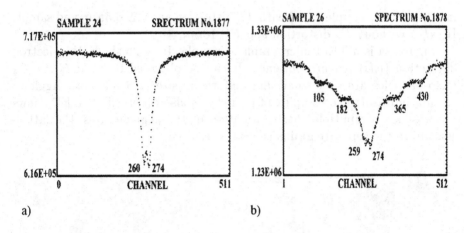

Figure 3. Moessbauer spectra of Cu-ferrite powders (a) sI (sample 24) with PS 6.5 nm and (b) sII (sample 26) with PS 10 nm.

magnetization of the nanoparticle sample with JT distortion exhibits a high-field irreversibility. The TEM data indicates grain agglomeration in the Cu-ferrite powders - this is the reason why we performed two types of experiments at high magnetic field and room temperature: we measured dry powders immobilized in parafin and free particles filling the sample holder. No noticeable difference can be observed one can, therefore, assume that the aggregation does not influence strongly the high-field interactions in the powder.

The $M(H)$ curves of the two samples at 4.2 K are compared in fig.5. The shape of the curve for $CuFe_2O_4$ without JT distortion exhibits a more abrupt onset in the high-field region and seems quasi saturated after 40 kOe. The hysteresis loop is open in the whole range of field variation with the positive and negative field sweeps being well separated, up to 140 kOe for the sample without JT distortion.

3.2. DISCUSSION

It is known that applying sufficiently high magnetic fields to ferrimagnetics systems can induce non-collinearity in the sublattice magnetic moments arrangement. The material reaches its saturation magnetization at very high magnetic fields [12]. If a high magnetic field H is applied to a ferrite with two magnetic sublattices, the energy of the system is expressed as follows:

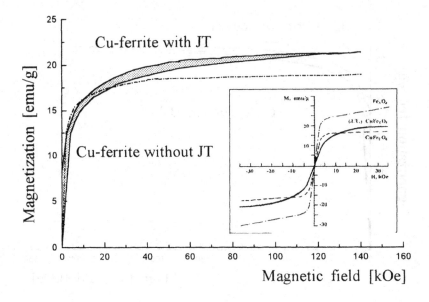

Figure 4. $M(H)$ curves at 295 K of nanoparticles of Cu-ferrite samples; sI with JT distortion (solid curve with magnetic losses) and sII (dashed curve) without-hysteresis curve. Inset: Lower field without-hysteresis curve of ultrafine powders of sI and sII compared with nanosized Fe_3O_4

$$E = -H(\mathbf{M_1} + \mathbf{M_2}) + J_{12}\mathbf{M_1M_2} + J_1\mathbf{M_1^2} + J_2\mathbf{M_2^2} \qquad (1)$$

where $\mathbf{M_1}$ and $\mathbf{M_2}$ are the magnetic moments of the sublattices while the J_i's are the exchange integrals in the sublattices, i.e., J_1 and J_2 in sublattice 1 and 2 respectively and J_{12} for the sublattices interaction.

At $H < H_{c1}$ (where H_{c1} is the first critical field), the ferrite has a collinear Neel structure. The volume energy of the magnetic interactions between the magnetic sublattices is $E_{\uparrow\downarrow}$ when the magnetic moments of both sublattices are antiparallel. The total magnetic moment of the system is $\mathbf{M_\Sigma} = \mathbf{M_1} - \mathbf{M_2}$.

When the magnetic field energy is higher than the volume interaction energy, the collinear structure changes into a non collinear one, (the so called "paraprocess"); in this case the magnetic moments of the sublattices begin to align with the direction of the applied external field, respectively at angles θ_1 and θ_2, different from zero, where $\mathbf{M_1}\, sin(\theta_1) - \mathbf{M_2}\, sin(\theta_1) = 0$.

When the critical field H_{c2} is reached, the system regains collinearity once more and $\mathbf{M_\Sigma} = \mathbf{M_1} + \mathbf{M_2}$, the total energy is $E_{\uparrow\uparrow}$ and the orientation

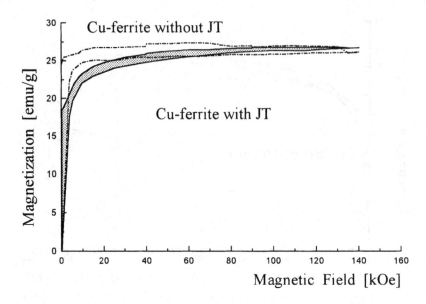

Figure 5. M(H) curves at 4.2 K of nanoparticles of Cu-ferrite samples; sI (solid curve) and sII (dashed curve)

of the moments in both lattices becomes parallel. The slope of the curves $M(H)$ between the two critical fields H_{c1} and H_{c2} is determined by the magnetic susceptibility according to the formula [12]

$$\chi = \mathbf{M_1}\mathbf{M_2}/E_{12} = (\mathbf{N_1}\mathbf{N_2})^{(1/2)}\mu_0^2/E_{12}. \qquad (2)$$

As one can see in figures 4 and 5, the particles with and without JT distortion differ in the values of χ (the slope of the $M(H)$ curve) at 4.2 K, which in presence of the same quantity of iron cations can be due to other magnetic interactions, these we associate with local changes of J_{eff} (effective exchange integral) as a result of the JT distortion.

In the real $CuFe_2O_4$, the JT cooperative effect distorts the octahedral state by stretching the axial copper-oxygen bonds to 0.24 nm and reducing the other four bonds to 0.19 nm, thus creating the square-planar geometry of the copper with the four nearest oxygens. Cu^{2+} ions have no strongly expressed inclination for octahedral coordination so octahedral and tetrahedral positions exist for the copper in the spinels. Tetrahedral deformation and $c/a > 1$ had been observed in the Cu-ferrospinels investigated. The distortion of the lattice in the area of Cu^{2+} can change the local symmetry of the crystal and noticeably influence the magnetocrystalline anisotropy. This mechanism can be much more complicated than the simple change of the constant of anisotropy; in this case areas of raised concentration of deforming ions can be formed, which results in reducing the total crystal's energy. Goodenough had proposed a similar hypothesis [13]. The presence of non-equivalent sublattices determines the existence of a variety of exchange interactions.

The nanocrystal size is comparable to the critical lengths that govern the magnetic and structural phenomena in the materials. If more than one type of barriers are present, the shortest characteristic length dominates the magnetic properties. For example, the magnetization cannot follow the random orientation of the single grain once the grain diameter is smaller than the length of ferromagnetic exchange interaction, which is about 30 nm [11] and this is the reason for the change in magnetization in comparison with the bulk one. In the absence of a magnetic field, the energy of a particle is minimized when its magnetic moment coincides with the anisotropy axis. Single-domain particles behave as large paramagnetic moments and exhibit superparamagnetism. The fundamental magnetic lengths are the crystalline anisotropy length L_K, and the magnetostatic length L_S. For the ferrospinels, these lengths are on the order of $L_K \approx 100$ nm and $L_S \approx 10$ nm. The latest investigations on nanosized ferrioxides showed that both the saturation magnetization and the high-field irreversibility are strongly dependent on the particle size. In the most common hypothesis [7], the phenomenon of high-magnetic field irreversibility in $M(H)$ is related to the magnetostatic length and, especially below 50 K, with the onset of the freezing process of the surface spin-glass layer. The open hysteresis loop at high magnetic field are explained as being the result of irreversible changes between these surface spin configurations and the ground state.

In the spinel, the two nonequivalent and opposite orientations of the sublattice magnetic moments are separated by an energy barrier, which is in the same order of magnitude, or lower, than L_S. From the point of view of superexchange interactions, in this case the most favorable length of metal-oxygen-metal bond is $2D < 0.48$ nm.

We associate the presence of irreversible magnetic losses in the case of a JT cooperative effect with the particularities of superexchange interactions in the spinel at induced non-collinearity in fields higher than the magnetic exchange interactions between the magnetic sublattices. These interactions appear at a given critical volume of the particle, where the length M-O-M (M-magnetic cation, O-oxygen) plays a predominant role. Following Andersons scheme [14], the ions have one $d(f)$-electron with non-compensated magnetic spin moment, while the O ion has p-electrons. If we take under consideration the pure ionic state of the ferrospinel, "exchange" magnetic interactions can appear in O-ion excited state; one p-electron of the oxygen moves then to one of the M-ions M and an M-O bond appears. Then the second electron of O can participate in exchange interactions with the other neighboring M-ion; one can then speak of an effective exchange connection between the M-ion spins in the M-O-M bond. In the scheme here discussed, the effective exchange connection of an "electronic" magnetic atom has the form [14]

$$H_{d1d2} = \pm b^2 I_{pd}(S_{d1}S_{d2}) \equiv J_{eff} \qquad (3)$$

where b is determined by the transport integral and the orbital states and does not depend on the spin changes; I_{pd} is the exchange integral between the electrons of the non-magnetic ions being in p-state and $d(f)$-electrons of the magnetic ions; S_{d1}, S_{d2} are the spin magnetic moments of the magnetic electrons; J_{eff} is the effective exchange integral. To solve the problem of the singlet and triplet energy state is thus reduced to:

$$J_{eff} = [-4\rho^2/(E_0 - E_1)^3][J_{pd}(J_{d'd} - J_{pd})] \qquad (4)$$

where E_0 is the electron self-energy, E_1 is the energy of the excited state and ρ is Coulomb distribution energy. The Kramers-Anderson formula is derived under the assumption that the interatom exchange integral $J_{d'd}$ (d - half-filled, d'- not filled d-state) is much low than the difference $E_0 - E_1$. In reality, however, these quantities have comparable values. In this case, one should consider two excited states with opposite signs, where $E_{\uparrow\downarrow} = E_i$-$(E_1 + J_{dd'})$ is the rise of the energy due to electron transfer in the d' state where the spin orientation in the $d(f)$-orbit is antiparallel, and $E_{\uparrow\uparrow} = E_0$-$(E_1 - J_{dd})$, where the spin orientation in d is parallel. Thus

$$J_{eff} = \rho^2 J_{pd}(1/E^2_{\uparrow\downarrow} - 1/E^2_{\uparrow\uparrow}). \qquad (5)$$

For a crystal unit cell of the ferrospinel with normal ion chemical connection $(d_{x^2-y^2})$, spin saturation and $J_{pd} < 0$ and $J_{eff} > 0$ are assumed; this is the reason for the parallel spin orientation in the separate sublattices and the antiferromagnetic exchange interactions between the sublattices, which Anderson connects with kinetic system state. In the case of JT distortion (d_{z^2}), a local energy minimization of the system is possible because of the oxygen polarization, when JT ion has the same spins orientation of the p-electrons, where the antiferromagnetic interaction between these ions is stabilized and a local non-collinearity is possible between the magnetic moments in one of the sublattices; these interactions arise when energy is added by an external magnetic field.

At temperatures near the absolute zero and external fields higher than the degree of the exchange interaction, the particle is paramagnetic and its total magnetic moment is $M_\Sigma = M_A + M_B$ as a result of the paraprocess.

If one assumes that for a certain critical size of the particle the superexchange coherence length (λ_{SE}) is dominant, the two equivalent but opposite orientations of the two sublattices magnetic moments are separated by an energy barrier U. As the temperature rises, the thermal transitions dominate over the energy barriers. The rate of this process is described by $exp(-U_{A-B}/k_B T_1)$, where U_{A-B} is the energy barrier of the superexchange interaction between A and B sublattices and $k_B T_1$ the thermal energy (k_B is the Boltzmann constant) and T_1 corresponds to the temperature at which the thermal fluctuation dominate over the U_{A-B} energy barrier. The particularities in the superexchange interaction in the ferrospinel are the reason for the energy barrier of the superexchange interaction above a certain temperature to be determined by the A-A and B-B interactions, because of the lower coherence length and $exp(-U_{A-A}/k_B T_2)$ or $exp(-U_{B-B}/k_B T_2)$ and $T_2 >> T_1$. When $T > T_1$ rises, λ_{SE} becomes predominant. When in one of the magnetic sublattices the magnetic symmetry is disturbed, one assumes that paraprocesses localized in one of the lattices are possible, - their existence being temperature-dependent. On the other hand, non-collinear magnetic ordering cannot exist simultaneously in the two sublattices of the perovskite structure [11]; if it arises in one sublattice, then the magnetic moments in the other will keep their collinearity.

That gives us the reason to suppose that $E_{A-A}(T) \neq E_{B-B}(T)$ and that at a given temperature it is possible that the two sublattices appear as different magnetic phases where for the one sublattice $E_{\uparrow\downarrow}$ and for the other $E_{\uparrow\uparrow}$ are minima.

While in high magnetic fields at low temperature the non compensated magnetic moments at the particle surface have an important role for the magnetic losses, the magnetic losses at higher temperatures (see Figure 4) are probably related to the exchange interactions in the magnetic sublattices of the ferrite below some critical size. For Cu-ferrite with JT distortion, a possibility appears to reach a stable state with parallel magnetic moments between the A and B sublattices not due to the rise of collinearity, but rather due to the paraprocesses in the octahedral sublattice where for a certain volume of the particle and for $T > T_1$ the area where the process exists can be considered as an independent phase. On the other hand, in magnetic microcrystals with particle dimensions below 20 nm, the magnetic anisotropy energy becomes comparable to the thermal energy even below room temperature, so that the magnetization vector fluctuates and the phase with paraprocesses behavior will have different $M_{A-A}((\tau T) \neq M_{B-B}(\tau T)$ dependence. It is obvious that in real systems one has to take into account the fact that the exchange interactions are the result of the simultaneous actions of many mechanisms and the exact answer can be given when the changes of the exchange integral of each of the two sublattices are considered.

4. Conclusion

The open hysteresis curve up to 140 kOe of nanosized Cu-ferrite particles belong to the area of the paraprocesses, i. e. when the technical saturation is completed and noncollinear orientation in the directions of the magnetic moments appears under the influence of an external magnetic field. At temperatures near the absolute zero, the freezing of non-compensated magnetic moments at the particle surface play an important role for the magnetic losses, while at higher temperatures these losses are probably related to the exchange interactions and paraprocesses in the magnetic sublattices of the ferrite. The paraprocesses of the ferrimagnetic oxides are determined by the length of superexchange interactions of the magnetic ions through the oxygen and are different from the exchange interactions in the metals or spin-glasses. If in the ferrioxide crystalline unit cell there exist local distortions of the sublattices or inter-, intralattices exchange interactions, a large number of disoriented thermal spin movements can be present. In the case of a cooperative JT effect, a low effective exchange field H_{eff} acts, whose direction coincides with that of the external magnetic field. At higher magnetic fields, and up to some critical temperature, the possibility to reach a stable state of the system with parallel magnetic moments between the A and B sublattices appears as a result of the induced noncollinearity in one of the sublattices - this a temperature-dependant process. We believe that up to some critical grain-size in this case both sublattices interact as independent phases with different magnetic orientation and the exchange interactions between these phases depend on the temperature.

5. Acknowledgements

The author wishes to thank to Dr. Y. Klamut and Dr. T. Palewski from the Laboratory of High Magnetic Fields and Low Temperatures for providing the opportunity for high magnetic field measurements. This work has been supported by NATO Grant HTECH.L6.973786 and Bulgarian National Fund for Scientific Research Grant No TH947.

(*) The high-magnetic field experiments were performed at the Laboratory of High-Magnetic Fields and Low Temperatures, 95, Gajowska Str., 53-529 Wroclaw, Poland

References

1. Tailhades, P., Bouet, L., Presmanes, L. and Rousset, A. (1997) Thin films and fine powders of ferrites: Materials for magneto-optical recording media, *J.Phys. IV France* **7**, C1-691-694.
2. Fannin, P.C. (1998) Wideband measurement and analysis techniques for the determination of the frequency-dependent, complex susceptibility of magnetic fluids, *Adv. Chem. Phys.* **104**, 181.
3. Chan, D.C.F., Kirpotin, D.B. and Bunn, P.A. Jr. (1993) Synthesis and evolution of colloidal magnetic iron oxides ... , *J. Magn. Magn. Mater.* **122**, 374-378.
4. Kodama, R.H., Berkowitz, A.E., McNiff, E.J., Jr., and Foner, S. (1996) Surface spin disorder in $NiFe_2O_4$ nanoparticles, *Phys. Rev. Lett.* **77**, 394-397.
5. Martinez, B., Obradors, X., Balcells, L., Rouanet, A. and Monty, C. (1998) Low temperature surface spin-glass transition in $\gamma - Fe_2O_3$ nanoparticles, *Phys. Rev. Lett.* **80**, 181-184.
6. Kodama, R.H., Makhlouf A. Salah, Berkowitz, A.E. (1997) Finite size effects in antiferromagnetic NiO nanoparticles, *Phys. Rev. Lett.* **79**, 1393-1396.
7. Kodama, R.H. and Berkowitz, A.E. (1999) Atomic-scale magnetic modeling of oxide nanoparticles, *Phys. Rev. B* **59**, 6321-6336.
8. Tronc, E. and Jolivet, J.P. (1992) *Magnetic Properties of Fine Particles* (eds. J-L.Dormann, D.Fiorani) 199.
9. Jolivet, J.P., Chaneac, C., Prene, P., Vayssieres, L. and Tronc, E. (1997) Wet chemistry of spinel iron oxides particles, *J. Phys. IV France* **7**, CI-573-....
10. Mitov, I., Cherkezova-Zheleva, Z. and Mitov, V. (1997) Comparative study of the mechanochemical activation of magnetite Fe_3O_4 and maghemite $\gamma - Fe_2O_3$, *Phys. Stat. Sol. (a)* **161**, 475-482
11. Nedkov, I., Merodiiska, T., Milenova, L. and Koutzarova, T. (1999) Modified ferrite plating of Fe_3O_4 and $CuFe_2O_4$ thin films, European Materials Research Society symposia, 1-4 June, Strasbourg, France, Abstracts GXP6 (to be published in *J.Magn.Magn.Mater.*)
12. Krupicka, V. (1973) *Physik der ferrite und der verwandten magnetischen oxide*, Prague; Belov, V. (1972) *Ferrites in the high magnetic fields*, Science, Russia
13. Goodenough, J.B., (1965) Chemical inhomogeneities and square BH loops, *J. Appl. Phys.* **36**, 2342-2374.
14. Anderson, P. W., (1950) Antiferromagnetism. Theory of superexchange interactions, *Phys. Rev.* **79**, 350-405.
15. Tejada, J., Ziolo, R.F. and Zang, X.X. (1996) Quantum tunneling of magnetization in nanostructured materials, *Chem.Mater.* **8**, 1784-1792 .

HIGH PERFORMANCES OXIDE ION CONDUCTORS
WITH COVALENT FRAMEWORK AND LIQUID-LIKE IONIC NETWORK

R.N. VANNIER, J.C. BOIVIN, G. MAIRESSE
Laboratoire de Cristallochimie et Physicochimie du Solide (L.C.P.S.)
UPRESA –CNRS 8012
Ecole Nationale Supérieure de Chimie de Lille
B.P. 108, 59652 Villeneuve d'Ascq Cedex, France

ABSTRACT. Depending on their structural dimensionality, bismuth-based oxide ion conductors can be classified in three groups : Bi_2O_3 itself is the prototype of the three-dimensional oxide ion conductors ; the BIMEVOX family, which particularly generated an increasing interest during the last ten years, belongs to the two-dimensional oxide ion conductors while $Bi_{26}Mo_{10}O_{69}$ and its related phases, with a structure based on $[Bi_{12}O_{14}]_\infty$ columns, can be considered as the first example of mono-dimensional bismuth-based oxide ion conductors. The common feature between these three classes of materials is that oxygen diffusion takes place in a very disordered network, exhibiting a liquid-like structure. The mono and two-dimensional oxide ion conductors have a larger stability range in temperature, since their structure is reinforced by a covalent skeleton. The main feature of the different structural characteristics of these groups of materials are presented and discussed.

1. Introduction

Oxide ion conductors are subject to considerable interest, since they are involved in the development of many new devices like oxygen sensors, Solid Oxide Fuel Cells (SOFC), dense oxygen separation membranes, catalytic membrane reactors, ...
Most of these devices (oxygen sensors, SOFC) are actually designed using Yttria-Stabilised Zirconia (YSZ). This electrolyte displays a high chemical stability both toward reductive atmosphere (H_2 or CO in SOFC cathode) and oxidising atmosphere (air or pure O_2). However, its limited oxide ion conductivity needs to operate at elevated temperature (800-1000°C) and such high temperatures lead to many technological problems : mechanical instability, materials ageing, ... Therefore, the challenge is to find new materials operating at temperature lower than 700°C, so that ancillary components in the systems could be fabricated from relatively cheap stainless steel alloys.
During the last decade, the discovery of new families of solid oxide electrolytes, exhibiting very high oxygen mobility in the temperature range 300-600°C, has opened new possibilities. Among them, the BIMEVOX, with the general formula $Bi_2V_{1-x}ME_xO_z$, where ME represents a large number of elements of the periodic table, exhibit the highest oxide ion conductivity [1]. Due to the sensitivity of Bi^{III} to reduction,

R. Cloots et al. (eds.), Supermaterials, 129–136.
© 2000 *Kluwer Academic Publishers. Printed in the Netherlands.*

these materials cannot be easily used in SOFC, but are particularly suitable for electrochemical separation of oxygen from air at moderate temperature 400-600°C [2]. Their exceptionally high conductivity is due to specific structural features, mainly related to the presence of the Bi^{III} cations. The stereochemically active non bonding electronic pair of this cation is responsible for very loose structures, which are particularly suitable for oxide ion migration. A further and determining advantage of bismuth is its catalytic effect on the oxygen dissociation, which is always the first and often the limiting step in every electrochemical process involving oxygen transfer.

From a structural point of view, the BIMEVOX materials can be considered as two-dimensional oxide ion conductors. They derive from the parent phase $Bi_4V_2O_{11}$, whose structure is built up from strongly covalent $Bi_2O_2^{2+}$ layers alternating with highly disordered vanadium oxygen deficient ionic sheets.

The same structural characteristics : a covalent framework and a liquid-like ionic network, are found in a new class of oxide ion conductors whose structure is built upon highly bonded $[Bi_{12}O_{14}]_\infty$ columns, connected to an ionic area [3,4].

The purpose of this paper is to underline the common structural features of these two classes of materials and to compare them with the three-dimensional bismuth-based oxide ion conductors, whose Bi_2O_3 itself is the prototype.

2. The BIMEVOX : two dimensional bismuth-based oxide ion conductors

The BIMEVOX materials derive from $Bi_4V_2O_{11}$ by partial substitution for vanadium with a metal. Depending on the temperature, $Bi_4V_2O_{11}$ exhibits three polymorphs. The α form is stable from room temperature up to about 430°C, the β form, between 430 and 570°C, and the γ one, above 570°C.

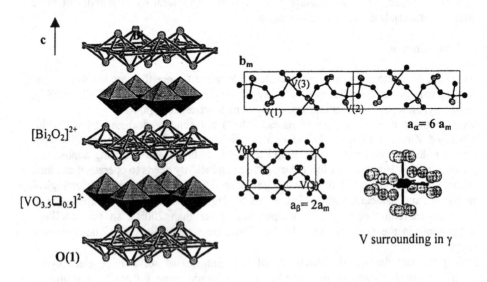

Figure 1. $Bi_4V_2O_{11}$ idealised structure, α and β $[VO_{3.5}\square_{0.5}]^{2-}$ sheets and γ vanadium surrounding

From a structural point of view, these three polymorphs can be described using a common orthorhombic mean cell with approximate parameters $a_m \approx 5.5$Å, $b_m \approx 5.6$Å, $c_m \approx 15.3$Å. The average structure consists of a packing of $[Bi_2O_2]^{2+}$ layers sandwiched between $[VO_{3.5}\square_{0.5}]^{2-}$ oxygen deficient perovskite sheets (Figure 1.). The presence of intrinsic oxygen vacancies confers to this material its exceptional electrical performances (Figure 2.).

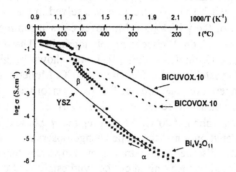

Figure 2. $Bi_4V_2O_{11}$ and BIMEVOX conductivity values versus temperature compared with YSZ

The ability for vanadium to accommodate various oxygen environments : tetrahedral, bipyramidal, octahedral, ... explains the singularity and the complexity of these materials. The structures of the different polymorphs have been refined using both X-ray and neutron diffraction data [5]. The β form is the most ordered one, it is also the polymorph exhibiting the highest activation energy. Its vanadium sheets consist of corner sharing tetrahedral and triangular bipyramidal sites along [1 0 0], leading to a $2xa_m$ superstructure (Figure 1.). A $6xa_m$ superstructure is observed for the α polymorph, resulting in an ordering of tetrahedral, double triangular bipyramidal sharing one edge, square pyramidal and octahedral oxygen polyhedra. However a higher disorder is observed around V(3) sites. The oxygen diffusion seems to occur through the oxide ions located around these sites. However, as only a few oxide sites are involved in the conduction process, this results in a low conductivity value.

The most conductive form is the γ one, which is also the most disordered one. Its structure has been refined in the I4/mmm space group [1,5] and looks like the idealised one. However, to take into account all the nucleon and electronic densities, all the atom sites were splitted, except O(1), which constitutes the basis of the $[Bi_2O_2]^{2+}$ sheets. A representation of the vanadium surrounding in this polymorph is given in Figure 1.. From a static point of view, this highly disordered oxygen surrounding could correspond to a projection of different oxygen environments : tetrahedral, octahedral, bipyramidal. However, the maps of nucleon density around vanadium reveal a large spread, better described as a liquid-like domain in which the oxygen diffusion takes place.

The common feature between the three polymorphs is the great stability of O(1) site that displays small isotropic temperature factors and short distances with bismuth atoms, typically 2.1-2.4 Å. It reveals the covalent character of the $[Bi_2O_2]^{2+}$ layers which clearly constitute the skeleton of the structure.

This γ form can be stabilised at room temperature by partial substitution for vanadium with numerous metals [6]. The BICUVOX were the first evidenced BIMEVOX. The electrical properties of BICUVOX.10, obtained by partial substitution for vanadium with 10% of copper, are reported in Figure 2. as well as that of BICOVOX.10 and are compared to those of YSZ. Conductivity values similar to those of YSZ at 700°C are obtained at 300°C. These value extrapolates that of γ $Bi_4V_2O_{11}$ in accord with the stabilisation at room temperature of this γ polymorph. However a slope change is to be noticed around 400°C. BICUVOX.10 structure was solved in the I4/mmm space group [1]. It is similar to that of the γ $Bi_4V_2O_{11}$. Moreover extra incommensurate reflections with h≈±0.3 were observed on Weissenberg photographs. These modulations were confirmed on BICUVOX.12 single crystals [7]. It was shown by high temperature X-ray diffraction, that these satellites vanished above 510°C. Muller et al. confirmed these results by high temperature neutron diffraction performed on a powder of BICOVOX.15 [8]. Superstructure reflections were observed at room temperature and reversibly disappeared at about 500°C. Therefore, the γ phases, stabilised at room temperature, are modulated ones and the slope change observed on the Arrhenius plots corresponds to an order-disorder phase transition. However, these modulations are rather complex and strongly depending upon the synthesis conditions. Moreover, they were more easily evidenced when using neutron diffraction instead of X-ray diffraction, which seems to indicate that only oxide are involved in this ordering process. Muller et al. refined the structure of BICOVOX.15 in the mean cell, using single crystal neutron diffraction data [9]. As for γ $Bi_4V_2O_{11}$, they observed a large spread of nucleon density around vanadium site, characteristic of a liquid-like domain.

The structure of these materials is close to that of the n=1 member of the Aurivillius family of phases, with a $[Bi_2O_2]^{2+}[A_{n-1}B_nO_{3n+1}]^{2-}$ general formula, where Bi_2WO_6 is the first member (n=1). However, Bi_2WO_6 presents no oxygen vacancy in its perovskite layers and is a very poor oxide ion conductor. Two solid solutions formulated $Bi_2W_{1-x}(Nb,Ta)_xO_{6-x/2}$ with $0 < x < 0.15$ were obtained by partial substitution for tungsten with niobium or tantalum [10] leading to extrinsic oxide ion conductors. Their electrical performances are reported in Figure 3. and are compared to those of the related BIMEVOX.

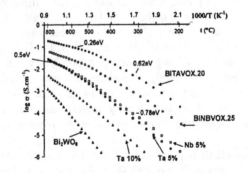

Figure 3. Bi2WO6 and its derivatives conductivity values versus temperature compared with related BIMEVOX

The best values were obtained for x=0.05, but they remain lower than those of the corresponding BIMEVOX. However, their Arrhenius plots exhibit a behaviour similar to that of the BIMEVOX with a slope change leading to two domains : a high temperature one and a low temperature one with activation energy of 0.5eV and about 0.8eV respectively.

The structure of these compounds is well defined and simple when compared with that of the BIMEVOX phases. Therefore, attempts to simulate oxygen diffusion in these materials by energy minimisation were undertaken, using Bi_2WO_6 as a model [11]. Migration energies were deduced for a range of oxygen vacancy jumps and revealed high barriers to oxygen ion migration within the $[Bi_2O_2]^{2+}$ layers (1.6eV) that confirmed the covalent character of these layers. High values were also obtained within the $[WO_6]$ equatorial plane (1.61-2.71eV). But, lower energies were observed for apical-equatorial jumps, suggesting a zigzag oxygen pathway through these sites, with activation energies of 0.45eV and 0.63eV, very close to the Arrhenius activation energy of 0.50eV observed at high temperature for the x=0.05 doped material. Formation of dopant-vacancy clusters was also considered and led to a binding energy of 0.25eV, in agreement with the higher experimental activation energy observed at low temperature. Therefore, the slope change in the conductivity could be related to dopant-vacancy associations at low temperature, which fully agrees an oxygen ordering (i.e. or vacancy ordering) at low temperature for the γ BIMEVOX phases.

Thereby, the γ BIMEVOX materials can be viewed as "semi-crystalline" phases, composed of a well crystallised part : the covalent $[Bi_2O_2]^{2+}$ layers, constituting the skeleton of the structure, alternating with a "vitreous zone" : the V-O sheets, in which the oxide diffusion takes place, greatly favoured by the bismuth $6s^2$ lone pair, oriented towards this liquid-like domain. The slope change in the conductivity is explained as a sudden softening of this part of the structure due to the sudden vanishing of some dopant-vacancy interactions, which can be compared to some glass transition.

3. $Bi_{26}Mo_{10}O_{69}$ and its derivatives : monodimensional bismuth-based oxide ion conductors

Our study of the BIMOVOX compounds, obtained by partial substitution for vanadium with molybdenum, enables us to evidence a new family of oxide anion conductors with a structure based on $[Bi_{12}O_{14}]_\infty$ columns.

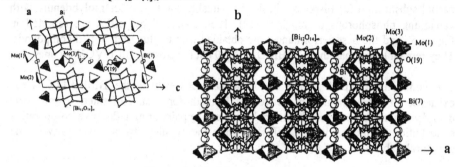

Figure 4. $Bi_{26}Mo_{10}O_{69}$ structure

134

It extends within a continuous solid solution domain in the ternary Bi_2O_3-MoO_3-V_2O_5 system around the composition $Bi_{26}Mo_{10}O_{69}$ [3,4]. The symmetry of $Bi_{26}Mo_{10}O_{69}$ is triclinic at room temperature and reversibly transforms to a monoclinic form at about 310°C. The increase of bismuth content or the introduction of at least 10% of vanadium in the molybdenum site leads to the stabilisation of the monoclinic form at room temperature. The extension of this solid solution towards the bismuth rich part of the Bi_2O_3-MoO_3 binary system is explained by the partial substitution for molybdenum with bismuth. $Bi_{26}Mo_{10}O_{69}$ and $Bi_{26}Mo_{9.6}Bi_{0.4}O_{68.4}$ crystal structures have been solved using single crystal X-ray diffraction data and powder neutron diffraction data, respectively. Similarly to the BIMEVOX materials, the structure of these phases can be compared with a "vitreous ceramic". It lies on well crystallised $[Bi_{12}O_{14}]_\infty$ columns, extending along [0 1 0], connected with a Bi-Mo-O ionic domain in which the mean square displacement of the oxygen atoms displays unusual high values likely correlated with some softness of this part of the lattice which favours the oxide ion mobility.

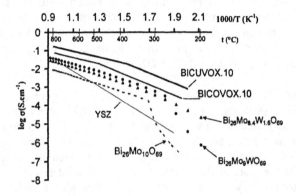

Figure 5. $Bi_{26}Mo_{10}O_{69}$ and its derivatives conductivity values versus temperature

These compounds are pure oxide ion conductors with an oxide diffusion favoured along [0 1 0] i.e. the axis of the columns, and can be considered as the first mono-dimensional bismuth-based oxide ion conductors. Their structural characteristics are preserved by partial substitution for bismuth with lead or alkaline earth, and for molybdenum with vanadium, phosphorous or tungsten [12]. Their electrical properties are reported in Figure 5. And the best performances are obtained with the tungsten doped compounds. They remain lower than those of BICUVOX.10 but are close to BICOVOX.10 ones.

The common structural feature of these new phases with the BIMEVOX is the existence of stereochemically active $6s^2$ bismuth lone pairs oriented towards the disordered ionic area. These highly polarisable lone pairs bring a decisive co-operative contribution to the oxide ion diffusion while the covalent frameworks provide the structural stability.

4. Comparison between mono, two and three-dimensional oxide ion conductors

As stipulated in the introduction, the first evidenced bismuth-based oxide ion conductor is Bi_2O_3 itself. Depending on the temperature, this compound exhibits two stable polymorphs : α and δ. Only the δ form is an oxide ion conductor and T. Takahashi, in 1970 [13], reported a conductivity value of $1 S.cm^{-1}$ at 730°C. The structure of the δ form is highly disordered but close to that of fluorite, with 25% intrinsic oxygen vacancies, responsible for its high oxide ion conductivity. However, this form is only stable in a narrow temperature domain, between a phase transition at 729°C and its melting point at about 830°C. The thermal effect associated with the $\alpha \leftrightarrow \delta$ transition is nearly three times larger than the heat of melting and corresponds to a true melting of the oxygen sublattice.

Another example of three-dimensional oxide ion conductor is the β-type phase in the Bi_2O_3-PbO system which exhibits an anti α-AgI crystal structure. A value of 1 $S.cm^{-1}$ was obtained for $Bi_{1.15}Pb_{0.85}O_{2.57}$ at 580°C. This structural type was also evidenced in Bi_2O_3-CdO, Bi_2O_3-CdO-PbO, Bi_2O_3-Sb$_2$O$_3$-PbO and Bi_2O_3-Ln$_2$O$_3$-PbO (Ln=Gd, Dy, Tm) systems. These phases are pure oxide ion conductors but also in a limited temperature range. The highly disordered liquid-like 3-dimensional oxygen network, whose formation is associated with a strong endothermic effect, necessary happens very near the melting point of the sample (630°C for $Bi_{1.15}Pb_{0.85}O_{2.57}$) [14].

All the chemical bonds in these materials are mainly ionic, leading to very high oxide ion mobility. Because of the large anion disorder, low interactions between anion and cation in the network occur and a poor stability as a function of the temperature is the counterpart. On the contrary, the presence of a covalent framework in 1D and 2D conductors results in a slight lowering of the conductivity value but significantly increases the temperature range of stability of the conductive form.

TABLE 1. summarises some of the domains of thermal stability and the conductivity values at transition temperature of these different types of oxide ion conductors.

TABLE 1. Thermal stability domain and dimensionality of some bismuth-based oxide ion conductors

	σ value at the transition temperature	Stability range	Dimensionality
δ Bi$_2$O$_3$	1 S.cm^{-1} at 730°C	100°C	3D
β Bi$_{1.15}$Pb$_{0.85}$O$_{2.57}$	1 S..cm^{-1} at 580°C	50°C	3D
γ Bi$_4$V$_2$O$_{11}$	0.15 S.cm^{-1} at 550°C	280°C	2D
Bi$_{26}$Mo$_{10}$O$_{69}$	0.3 mS.cm^{-1} at 310°C	550°C	1D

The best conductivity values are observed for three-dimensional oxide ion conductors, but in a narrow temperature domain in which precludes their use for practical applications.

5. Conclusion

The common feature between these three classes of materials is that oxygen diffusion takes place in a very disordered network, exhibiting a liquid-like structure. The main difference is their temperature range of stability. Three dimensional materials,

136

possessing no "permanent" skeleton, are only stable within a limited temperature range, typically lower than one hundred of degrees. This range quickly increases as the dimensionality of the structure decreases. As a consequence, a two dimensional structure appears as a good compromise between high oxide mobility and large domain of thermal stability.

6. References

1. Abraham, F., Boivin, J.C., Mairesse, G. and Nowogrocki, G. (1990) The BIMEVOX series : a new family of high performances oxide ion conductors, *Solid State Ionics* **40/41**, 934-937

2. Boivin, J.C., Pirovano, C., Nowogrocki, G., Mairesse, G., Labrune, P., Lagrange, G. (1998) Electrode-electrolyte BIMEVOX system for moderate temperature oxygen separation, *Solid State Ionics* **113-115**, 639-651

3. Vannier, R.N., Mairesse, G., Abraham, F. and Nowogrocki, G. (1996) $Bi_{26}Mo_{10}O_8$ solid solution type in the Bi_2O_3-MoO_3-V_2O_5 ternary diagram, *Journal of Solid State Chemistry* **122**, 394-406

4. Vannier, R.N., Abraham, F., Nowogrocki, G. and Mairesse, G. (1999) New structural and electrical data on Bi-Mo mixed oxides with a structure based on $[Bi_{12}O_{14}]_\infty$ columns, *Journal of solid State Chemistry* **142**, 294-304

5. Pernot E. (1994) Relation entre synthèse, structure et conductivité dans $Bi_4V_2O_{11}$ et ses substitués (Cu, Ni), Thesis, University of Grenoble

6. Mairesse, G. (1993) Bismuth-based oxide conductors, novel structural and electrical features, *Fast Ion Transport in Solids*, 271-290, Kluwer Academic Publishers, the Netherlands

7. Pernot, E., Anne, M., Bacmann, M., Strobel, P., Fouletier, J., Vannier, R.N., Mairese, G., Abraham, F., Nowogrocki G. (1994) Structure and conductivity of Cu and Ni-substituted $Bi_4V_2O_{11}$ compounds, *Solid State Ionics* **70/71**, 259-263

8. Muller, C., Anne, M., Bacmann, M. (1998) Lattice vibrations and order-disorder transition in the oxide anion conductor BICOVOX.15 : a neutron thermodiffractometry study, *Solid State Ionics* **111**, 27-36

9.   Muller, C., Anne, M., Bacmann, M. and Bonnet, M. (1998) Structural studies of the fast oxygen ion conductor BICOVOX.15 by single-crystal neutron diffraction at room temperature, *Journal of Solid state Chemistry* **141**, 241-247

10. Baux, N., Vannier R.N., Mairesse, G., Nowogrocki G. (1996) Oxide ion conductivity in $Bi_2W_{1-x}ME_xO_{6-x/2}$ (ME=Nb, Ta), *Solid State Ionics* **91**, 243-248

11. Islam, M.S., Lazure, S., Vannier, R.N., Nowogrocki, G. and Mairesse, G. (1998) Structural and computational studies of Bi_2WO_6 based oxygen ion conductors, *Journal of Materials Chemistry* **8(3)**, 655-660

12. Vannier, R.N., Danzé, S., Nowogrocki, G., Huvé, M. and Mairesse, G., A new class of mono-dimensional bismuth-based oxide anion conductors with a structure based on $[Bi_{12}O_{14}]_\infty$ columns, *Solid State Ionics*, to be published

13. Takahashi, T., Iwahara, H., Nagai, Y. (1970), *J. Electrochem. Soc.* **117**, 244

14. Boivin, J.C., Mairesse, G. (1998) Recent materials developpements in fast oxide ion conductors, Chemistry of Materials 10, 2870-2888 and references therein

ELECTRONIC CRITERION FOR THE EFFECTIVENESS OF LITHIUM INTERCALATION IN TRANSITION METAL COMPOUNDS

J. MOLENDA
Department of Solid State Chemistry,
Stanislaw Staszic University of Mining and Metallurgy,
al. Mickiewicza 30, 30-059 Cracow, Poland

1. Introduction

Transition metal compounds with a general formula $A_xM_aX_b$ (A = alkaline metal, M = transition metal, X = O, S, Se) constitute a group of potential electrode materials for a new generation of $A/A^+/A_xM_aX_b$ batteries. This application is related to the fact that these compounds can reversibly intercalate high amounts of alkaline ions (1 or more moles per mole of M_aX_b) already at room temperature, without significant changes in their crystallographic structure. In a numerous group of M_aX_b compounds capable of intercalating foreign ions, particularly interesting are those having a layered or frame structure. The ionic transport in these systems is related to appreciable mobility of alkaline ions in two-dimensional interlayer spaces or tunnels whereas the electronic transport is related to *d* electrons. Investigations of $A_xM_aX_b$ intercalated compounds carried out so far in many laboratories in the world [1-3] are connected with their principal application, i.e. in the alkali cells. The investigations are concentrated on the search for electrode material of high energy efficiency. Investigations of the properties are limited to the studies of the crystallographic structure and its modification in the intercalation process and to the measurements of the chemical diffusion coefficient of lithium. Practically there is a shortage of basic research on the transport and electronic phenomena in intercalated electrode materials. The published useful parameters of cells frequently refer to the undefined initial structure of the ionic and electronic defects of the cathode material. The attempts made so far to explain the observed diverse electrochemical properties of $A_xM_aX_b$ compounds, based on their crystallographic structure, formation of a superstructure etc. are not satisfactory and do not lead to determination of a criterion of electrochemical properties of the systems.
The investigations on defect structure and transport properties (deviation from stoichiometry, electrical conductivity, thermoelectric power) carried out under the

R. Cloots et al. (eds.), Supermaterials, 137–151.
© 2000 *Kluwer Academic Publishers. Printed in the Netherlands.*

conditions of thermodynamic equilibrium at elevated temperatures enable to determine the ionic and electronic defect structure, charge transport mechanism and optimal thermodynamic parameters (T, pX_2) for the synthesis of electrode materials with desired transport properties.

2. Electronic model of intercalation process

The author of this work basing on numerous investigations on $A_xM_aX_b$ (A = Li; M = 3d, 4d, 5d; X= O, S, Se) has pointed out that the electronic structure of these materials plays an important role in the intercalation process [4-7]. The intercalation of alkaline metal into transition metal compounds showing metallic or semiconducting properties can be considered as a reversible topotactic redox reaction, in which the transition metal changes its valency. The insertion of alkaline metal A to the conducting M_aX_b lattice takes place by introduction of alkaline ions A^+ together with the equivalent number of electrons:

$$xA^+ + xe^- + M_aX_b \leftrightarrow A_xM_aX_b \qquad (1)$$

Variations of EMF of the $A/A^+/A_xM_aX_b$ cell, accompanying the intercalation reaction, correspond to those of chemical potential of electrons (Fermi level) of the cathode material brought about by alkaline metal doping. High density of states near the Fermi level is associated with wide range of alkaline metal concentration and weak compositional dependence of the cathode potential, whereas the existence of delocalized electronic states guarantees rapid relaxation of the cathode material and good useful properties. Figs.1-2 show the expected cathode potential variations in function of electronic structure of the cathode material, $A_xM_aX_b$. The character of state density function determines the shape of the discharge curve.

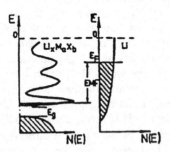

Fig.1 Density of states of $Li_xM_aX_b$ and lithium ilustrating chemical potential variations of electrons and related EMF of $Li/Li^+/ Li_xM_aX_b$ cell

The concept of the Mott localization makes possible to explain the two types (step-like and monotonous) of the discharge curve. Systems with $R_{M-M} > R_c$ correspond to the Mott insulators while those with $R_{M-M} < R_c$ represent correlated metals. Under such a classification scheme there is a principal difference between the two kinds of systems after intercalation. In the intercalated systems with $R_{M-M} > R_c$ the majority of

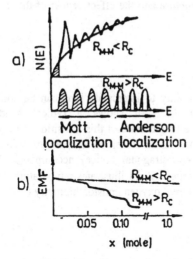

Fig.2 Electronic structure of the cathode material (a) and related changes of the EMF of the Li/Li$^+$/ Li$_x$M$_a$X$_b$ cell (b).

R_{M-M} - effective M-M distance
R_c - critical M-M distance for insulator-metal transition [8]

electrons are localized. In other words, only the number of electrons representing the deviation from the half-filled band situation (or holes) can be treated as carriers. On the contrary, in systems with $R_{M-M} < R_c$ all carriers in an uppermost partially filled band can be treated as carriers. In systems with $R_{M-M} > R_c$ the number of carriers is essentially smaller than in systems with $R_{M-M} < R_c$. This means that in the first case the electrons are influenced by the periodic potential of surrounding ions to much larger extent than in the latter case. In other words, the carriers in systems with $R_{M-M} < R_c$ can be regarded as nearly-free electrons with rather featureless density of states due to the screening of the periodic potential. This is not true for systems with $R_{M-M} > R_c$ in which the screening is not so effective and the density-of-states-curve is usually spiky with a rather complicated type of shape as a function of particle energy. This explains the monotonous character of the discharge curve as due to a monotonous variation of the Fermi energy and the large range of concentration changes of the alkali-ion component for systems with $R_{M-M} < R_c$, and a step-like character of the discharge curve and narrow range of alkali metal nonstoichiometry for systems with $R_{M-M} > R_c$. Such an approach gives explanation for both monotonous and step-like shape of the discharge curve and permits to design useful properties of the intercalated electrode materials.

It follows from the presented model that examination of the potential $A_xM_aX_b$ compounds by measuring the EMF of $A/A^+/A_xM_aX_b$ cells is an excellent tool of solid state physics which permits direct observation of the Fermi level changes during the doping process. The investigations of physico-chemical properties of the intercalated systems in function of alkaline metal concentration have an important cognitive aspect. Modification of electronic structure of the intercalated material permits to follow the relations among structure, composition, valency of transition metal, disorder and reactivity of solids which remains an open question in material science.

Below there will be presented selected electrode materials of extremely differing properties (insulator, metal, superconductor) with a view to demonstrate the existence of

140

close relation between the electronic structure and the effectiveness of the intercalation process.

3. Vanadium dioxide

Vanadium dioxide is a system which ability to intercalate ions can be controlled by modifying electronic structure. At 340K it shows a typical Mott-Hubbard semiconductor-metal transition [9]. The temperature of this transition is very sensitive to the presence of impurities and dopants. At the transition temperature VO_2 undergoes structural rearrangement (monoclinic $\leftrightarrow$ tetragonal lattice) accompanied by abrupt change of electrical and magnetic properties. Fig.3 illustrates temperature dependence of electrical conductivity of VO_{2-y} at different oxygen nonstoichiometry ($y_{III} > y_{II} > y_{I}$).

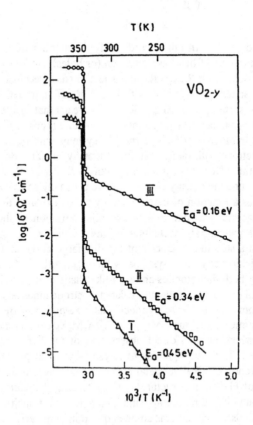

Fig.3 Temperature dependence of electrical conductivity for three VO_{2-y} series ($y_{III} > y_{II} > y_{I}$)

The observed discontinuity of electrical properties can be explained on the basis of a qualitative model of its electronic structure, shown in fig.4 [10]. Charge transport on the non-metallic side of the transition (T<T$_t$) proceeds via localized states. On the metallic side of transition the charge transport takes place in the effective band with carriers near the Fermi level however electronic gas is correlated.

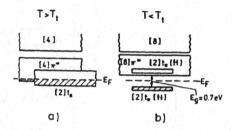

Fig.4 Qualitative scheme of the electronic structure of VO$_2$ [10]. (a) T>T$_t$, (b) T<T$_t$

Extensive studies [11] have shown that localization of electronic states is the reason why pure VO$_2$ does not accept lithium ions at room temperature (kinetic limit). It has been shown [11] that high purity vanadium dioxide which at room temperature did not undergo any intercalation at all, after doping with tungsten (due to weaker electronic correlation) acquired high capability of intercalating lithium ions what indicates that the electronic structure of the cathode material is decisive for the electrochemical behaviour.

4. Li$_x$CoO$_2$ system

Fig.5 presents a discharge curve of a Li/Li$^+$/Li$_x$CoO$_2$ cell in the OCV system. Points A, B, C, D, E denote compositions at which the work of the cathode material was stopped in order to examine its physical properties.

Fig.6 shows electrical conductivity of the cobalt oxide Li$_x$CoO$_2$ as a function of deintercalation degree. Comparison of figs. 5 and 6 confirms close correlation between the electronic and electrochemical properties of the cathode material. Cobalt bronze with compositions A and B (fig.5) situated within the potential jump exhibits semiconducting properties with energy gap corresponding to the height of the jump (0.7 eV) whereas at compositions situated on the potential plateau (points C, D, E in fig.5) it shows metallic properties. The analysis [6] indicates that good electrochemical properties of Li$_x$CoO$_2$ within the composition range 0.1<x$_{Li}$<0.97 are due to the existence of the effective energy band and delocalization of electronic states near the Fermi level.

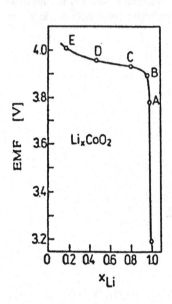

Fig.5 Charge curve EMF for Li_xCoO_2

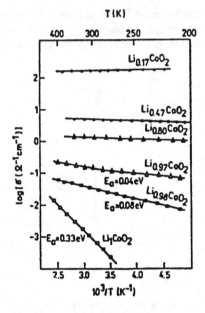

Fig.6 Temperature dependence of electrical
conductivity for Li_xCoO_2

5. $YBa_2Cu_3O_{7-\delta}$ superconductor

Pseudo-layered structure of the system Y-Ba-Cu-O and its high electrical conductivity seems to create favourable conditions for the intercalation process of lithium, and this was the reason to undertake investigations on this system with a view to examine its electrochemical properties.

In the system Y-Ba-Cu-O one can observe a reversible phase transition from the orthorhombic into tetragonal phase at the temperature 870-920K [12]. The orthorhombic phase is superconducting at low temperatures. Fig.7 shows the model of the electronic structure of nonstoichiometric $YBa_2Cu_3O_{7-\delta}$ at high temperatures, developed on the basis of the author's own investigations of the deviation from stoichiometry, electrical conductivity and thermoelectric power, carried out at high temperatures, under conditions of thermodynamic equilibrium, as a function of temperature and oxygen pressure. As follows from fig.7 for the orthorhombic phase with low nonstoichiometry δ one can observe an effective energy band of metallic properties. Increase of nonstoichiometry in the oxygen sublattice causes a narrowing of the effective energy band, accompanied by an increase in the effective mass of the carriers [7]. For the deviation $\delta = 0.7$ there occurs a splitting of the effective energy band into two subbands, in which the electronic states are localized.

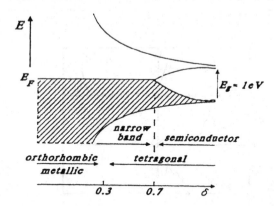

Fig.7 Scheme of the electronic structure of non-stoichiometric $YBa_2Cu_3O_{7-\delta}$ at high temperatures

As it has been found experimentally [7] only a freezing of the orthorhombic phase with metallic properties ($\delta < 0.3$) leads to superconducting properties at low temperatures. Identical criterion is true for the useful electrochemical properties. The system $YBa_2Cu_3O_{7-\delta}$ with $\delta > 0.7$ shows complete absence of reactivity in relation to lithium in the cell $Li/Li^+/Li_yYBa_2Cu_3O_{7-\delta}$ (the cell voltage drops rapidly to zero). This behaviour should be associated with the localization of the electronic states near the Fermi level (fig.7). The best electrochemical properties were obtained for $YBa_2Cu_3O_{7-\delta}$ with composition close to a stoichiometric one. Fig.8 shows the electromotive force of the cell $Li/Li^+/Li_yYBa_2Cu_3O_{6.96}$ as a function of lithium content in the cathode material.

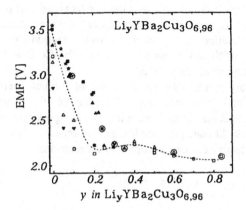

Fig.8 EMF of $Li/Li^+/Li_yYBa_2Cu_3O_{6.96}$ cell

144

The monotónous character of the discharge curve suggests a monotonic change in the position of Fermi level. The change in the position of Fermi level as a function of the concentration of carriers in the system Y-Ba-Cu-O can be traced on the diagram of the states density, determined by Mattheiss [13], (fig.9). Worth noting is the low density of states at Fermi level.

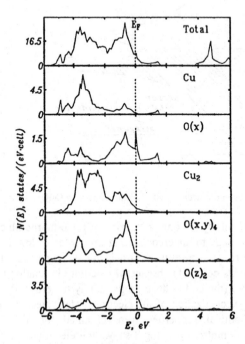

Fig.9 Calculated density of states for the $YBa_2Cu_3O_{7-\delta}$ system [13]

The electrons introduced in the intercalation process shift the position of Fermi level towards the tail of the band where the states are localized. This is confirmed by the activated character of electrical conductivity in the intercalated samples of $Li_yYBa_2Cu_3O_{7-\delta}$ (activation energy increases with increasing y_{Li}) [7]. The maximal amount of lithium which can be introduced into Y-Ba-Cu-O is 1.6 mole/mole, which makes it an interesting electrode material.

It appears interesting to compare the influence of lithium intercalation and oxygen nonstoichiometry on superconducting properties of the discussed system. The superconducting properties decay both with increasing oxygen nonstoichiometry and increasing content of lithium [7]. Both chemical reactions lead to a decrease in the concentration of effective carriers according to the reaction:

$$Cu^{+3} + e^- \leftrightarrow Cu^{+2} \tag{2}$$

An increase in the parameter a (a decrease of the overlap of Cu-O-Cu orbitals), both with the increase in lithium content and in oxygen nonstoichiometry, contributes to the decay of the superconducting properties [7].

6. Spinel $Li_{1-x}Mn_2O_4$

In manganese spinel $Li_1Mn_2O_4$ the conditions for charge transport are much worse. Its electrical conductivity at 300 K (i.e. in the conditions of the intercalation process) is very low, of the order of 10^{-4} Scm^{-1}. Charge transport at room temperature takes place through jumps of small polarons with a high activation energy (0.3-0.5 eV). Under such conditions the process should be extremely difficult. Nevertheless it proceeds with a good efficiency. So a question arises: what happens with this material during deintercalation? Detailed studies of transport properties combined with careful structural investigations [14] performed as a function of temperature for deintercalated spinel indicate a strong modification of its structure and electronic properties during deintercalation at T=300 K.

Fig.10 presents EMF of the $Li/Li^+/Li_{1-x}Mn_2O_4$ cell as a function of lithium content in a cathode material.

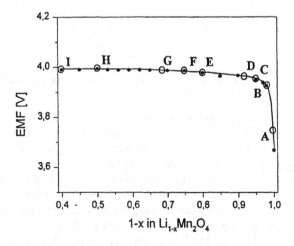

Fig.10 EMF of the $Li/Li^+/Li_{1-x}Mn_2O_4$ cell as a function of lithium content in cathode material. A, B... mark the compositions at which the work of a battery was stopped in order to study properties of the cathode material.

Extremely small changes of the $Li^+/Li_{1-x}Mn_2O_4$ cathode potential observed as a function of composition in the range $0.4 < 1-x < 0.96$ indicate a high density of electronic states at the Fermi level in the system under consideration. The process of electrochemical lithium deintercalation in manganese spinel may be expressed in the following way:

$$LiMn_2O_4 \leftrightarrow Li_{1-x}Mn_2O_4 + xLi^+ + xe^- \qquad (3)$$

Oxidation of manganese is the effective reaction [14]:

$$Mn^{+3} \leftrightarrow Mn^{+4} + e^- \qquad (4)$$

Fig.11 presents lithium chemical diffusion coefficient in $Li_{1-x}Mn_2O_4$ as a function of x_{Li}. As can be seen from the figure in the range $0.4 < 1-x < 1$ chemical diffusion coefficient of lithium increases with the degree of deintercalation and next it decreases.

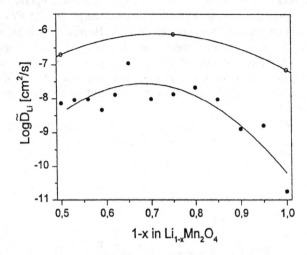

Fig.11 Lithium chemical diffusion coefficient in $Li_{1-x}Mn_2O_4$ obtained by galvanostatic intermittent titration technique. o - data obtained by Schoonman (1995)

A qualitatively different mechanism of charge transport was observed in the $Li_{1-x}Mn_2O_4$ system in comparison with the initial $Li_1Mn_2O_4$ spinel. Sharp maxima in termoelectric power and jumps of the conductivity were detected in the vicinity of room temperature (i.e. at working temperature of the battery, fig.12-15) what suggests a high density of states at the Fermi level. Detailed structural investigations of deintercalated spinel [14] revealed the occurrence of a broad phase transition (300-200 K) from cubic into orthorhombic structure. It seems that the observed maxima of thermoelectric power and conductivity jumps accompany this phase transition. During the transition areas of spatial ordering of Mn^{+3} and Mn^{+4} ions characterised by a high density of electronic states appear locally and they lead to the appearance of long-range interactions. The recent work of Rodrigez-Carvajal [15] on the nature of the phase transition in manganese spinel describes a specific type of ordering of the Mn^{+3} and Mn^{+4} ions (electronic crystallisation) which might explain the observed high density of states at the Fermi level and anomalous electronic effects. Good work of the $Li^+/Li_{1-x}Mn_2O_4$ cathode seems to be related to the existence of percolation path with a high density of states which is connected with ordering of the Mn^{+3} and Mn^{+4} ions. This ordering is present in a wide range of compositions x_{Li} and ensures good transport properties of electronic carriers and hence a high value of lithium chemical diffusion coefficient.

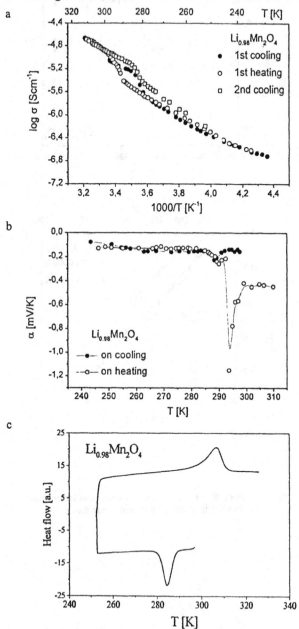

Fig.12 Electrical and thermal properties of $Li_{0.98}Mn_2O_4$ as a function of temperature
(a) electrical conductivity (b) thermoelectric power (c) DSC

148

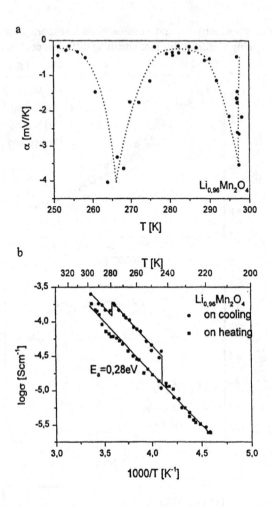

Fig.13 Electrical properties of $Li_{0.96}Mn_2O_4$ as a function of temperature
(a) thermoelectric power (b) electrical conductivity

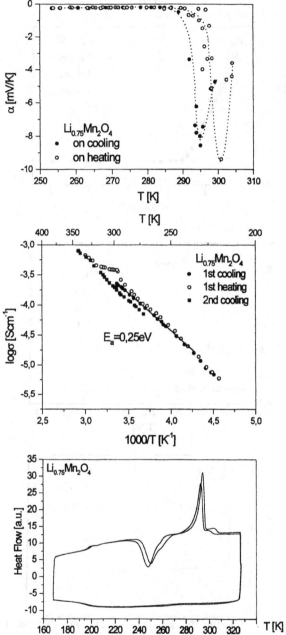

Fig.14 Electrical and thermal properties of $Li_{0.75}Mn_2O_4$ as a function of temperature
(a) thermoelectric power (b) electrical conductivity (c) DSC

150

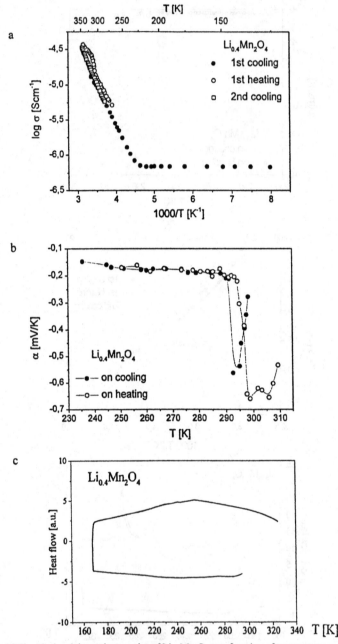

Fig.15 Electrical and thermal properties of $Li_{0.4}Mn_2O_4$ as a function of temperature
(a) thermoelectric power (b) electrical conductivity (c) DSC

References .

[1]   M.S. Whittingham, *J.Electrochem.Soc.* **123**(1976)315.
[2]   T.Ohzuku, K.Sawai, T.Hiraci, Denchi Gijutsu (Battery Technology) **3**(1991)14
[3]   T.Ohzuku *in* **Lithium Batteries. New materials, Developments and Perspectives** ed. G.Pistoia Elsevier 1994, p.239
[4]   J.Molenda, *Phys. Stat. Sol.(b)* **165**(1991)419
[5]   J.Molenda, A.Stoklosa, *Solid State Ionics*, **36**(1989)43
[6]   J.Molenda, A. Stoklosa, T.Bak, *Solid State Ionics*, **36**(1989)53
[7]   J.Molenda, T.Bak, A.Stoklosa, *Physica C*, **207**(1993)147
[8]   J.B. Goodenough, *Solid State Chem.*,**5**(1971)279
[9]   A.Zylbersztejn and N.F.Mott, *Phys.Rev.B* **11**(1975)4383
[10]   J.B.Goodenough, **Les oxydes des metaux de transition** (Gauthier-Villars, Paris, 1973)
[11]   J.Molenda, T.Bak, *Phys.Stat.Sol. (a)* **135**(1993)263
[12]   J.Molenda, A.Stoklosa, T.Bak, *Physica C* **175**(1991)555
[13]   L.F.Mattheiss, D.R.Hamann, *Solid State Commun.* **63**(1987)395
[14]   J.Molenda, K.Swierczek, M.Molenda, J.Marzec, J.Przewoznik, Cz.Kapusta, *Solid State Ionics* (in press)
[15]   J.Rodriguez-Carvajal, G.Rousse, C.Masquelier, M.Hervieu, *Phys.Rev.Letters* **81**(1998)4660

DYSPROSIUM SUBSTITUTION IN BI-BASED 2223 MATERIALS:
THE ROLE OF THE SUPERCONDUCTING LAYER CHARGE DISTRIBUTION FOR INDUCING A STRUCTURAL PHASE TRANSITION

R. CLOOTS, J. FELDMANN, A. RULMONT

SUPRAS, Chemistry Institute B6, University of Liège,
Sart-Tilman B-4000 Liège, Belgium

AND

M. AUSLOOS

SUPRAS and GRASP, Physics Institute B5, University of Liège,
Sart-Tilman B-4000 Liège, Belgium

Abstract. The chemical synthesis of dysprosium doped Bi-based 2223 high-T_c superconducting materials prepared by either a "two-powder process" or a "crystalline-glassy matrix precursor" method has been examined. In both cases, the 2223 phase is seen to be unstable. The reason for the Dy-doped 2223 instability is discussed and found to originate in electronically mediated bond-mismatch, resulting in a strain energy minimization by phase decomposition.

1. INTRODUCTION

Many experimental procedures have been recently developed in order to obtain suitable polycrystalline superconducting ceramic based materials characterized by good electrical transport properties like a sharp resistivity drop at the critical temperature T_c. Among these methods, the texturation process is undoubtlessly the most valuable way for obtaining high critical current density values in polycrystalline materials. Different techniques are useful, as for example melt-texturing processes in a thermal field, in presence of a magnetic field, a combination of both fields, or the seeding method. The second method is based on the magnetic anisotropy of the crystallo-

R. Cloots et al. (eds.), Supermaterials, 153–162.
© 2000 *Kluwer Academic Publishers. Printed in the Netherlands.*

graphic unit cell, this anisotropy being usefully enhanced by rare-earth ion (like dysprosium, holmium or erbium) substitution [1-5]. Such a process has been successfully used in order to produce well oriented dysprosium doped Bi-2212 so-called BSSCO materials [6-7]. We report in this paper the results following the same procedure to the 2223 phase by partially substituting the strontium and bismuth sites with dysprosium.

It is believed that the 2223 phase grows either from the 2212 phase via a dissolution-reprecipitation process [8-10] or through a bond breaking internal stress induced process [11-12]. In the first case, the factors governing the 2223 phase formation are the diffusion rates of the reactive species, i.e. the Ca^{2+} and Cu^{2+} ions. The phase conversion is thus kinetically limited. Many methods and precursors have been proposed to provide Ca^{2+} and Cu^{2+} ion reservoirs as fast pathways for species diffusion (for a rather extensive set of references see [13]). The second hypothesis requires fine tuning of the lattice spacing in the c-direction during an extraplane insertion. Both theories are a priori suggesting that the preparation of an "as pure as possible" dysprosium-doped Bi-based 2223 material is also feasible. However, only a few investigations on the effect of rare-earth substitution for calcium in the 2223 phase are reported in the literature [14-17]. These authors conclude that the 2223 phase is not stable for so-called "thermodynamic reasons". It is thus of fundamental interest to study the effect of a rare-earth ion like dysprosium i.e. a heterovalent substitution for strontium and homovalent for bismuth respectively on the thermodynamic stability of the 2223 phase, and to precise such reasons if possible, or even to discriminate between the physicochemical hypotheses recalled here above.

Two chemical procedures have been used in our work and are discussed in section 2, i.e. (i) a two-powder process via a solid state reaction for strontium-substituted Bi-2223, and (ii) the matrix method via a crystalline component and a glassy precursor made from two other components for bismuth-substituted Bi-2223 as discussed in section 2.1 and 2.2 respectively. The experimental procedure and the results from X-ray diffraction and Energy Dispersive X-ray microanalyses, and electrical transport measurements are given and discussed separately for each process in the following.

2. EXPERIMENTAL RESULTS

2.1. THE TWO-POWDER PROCESS

Dysprosium doped Bi-based 2223 nominal BSCCO ceramics compounds were aimed to be synthesized through a two-powder process by a solid state reaction occuring between appropriate amounts of $Bi_{2-x}Pb_xSr_{2-y}Dy_yCuO_z$ and $CaCuO_2$ phases for (i) x=0 and y=0; (ii) x=0 and y=0.2; (iii) x=0.3 and y=0 (sample FS9B); and (iv) x=0.3 and y=0.2 (sample FS11A). Only

TABLE 1. Parameters of the thermal cycle and the superconducting critical temperature measured from the electrical resistance curves for the Dy-doped (FS11A) and Dy-free (FS9B) Bi-based 2223 materials prepared by the two-powder process as described in the text.

T(°C)	T'(°C/h)	DT(days)	Intermediary grinding	T_c(K)	sample
845	150	6	1	112 - 78	FS9B
845	150	6	1	82	FS11A

the cases (iii) and (iv) are discussed here. Table I gives the parameters of the thermal cycle which was used for the preparation of these strontium substituted for dysprosium Bi-based 2223 materials. Figure 1 shows the X-ray diffraction patterns of samples FS9B and FS11A. In the latter, when dysprosium is added, only the 2212 phase is clearly visible.

The $Bi_{2-x}Pb_xSr_{2-y}Dy_yCuO_z$ compounds have been first synthesized by a classical solid state reaction from a stoichiometric mixture of the corresponding oxides and/or carbonates, heat treated at 740°C during 24 hours. The X-ray diffraction pattern shows that beside the peaks corresponding to the Bi-based 2201 phase some copper oxide secondary phases are clearly visible. It is important to note that not all Bi-free phases can be easily detected by X-ray diffraction analysis due to the greater X-ray scattering power of bismuth with respect to the other elements. For observing the Bi-free phases, an Energy Dispersive X-ray microanalysis has been performed on a polished section of the samples. This analysis has been performed with Bi_2CuO_4 and $SrCuO_2$ as internal standards. It is observed that strontium rich and copper rich precipitates can be emphasized. In fact, the composition of the Bi-phase resulting from the EDX microanalysis can be written as $Bi_{3+x'}(Sr,Dy)_{3-x'}Cu_{1+y'}O_6$ with $0 < x' < 0.5$ and y'=0.05 which is slightly copper deficient as compared to the classical 2201 superconducting phase.

Figure 2 gives the electrical resistivity as a function of temperature for samples FS9B and FS11A. Only for the Dy-free 2223 material (FS9B), a shift of the superconducting transition temperature from 112K to 75K is clearly noticed. For the Dy-doped material (FS11A), superconductivity appears only at 85K indicating the absence of 2223 superconducting phase along the percolation electrical pathway. This indicates that the dysprosium substitution at the strontium sites in the precursor inhibits the formation of the 2223 phase. These results confirm those already reported in the literature when dysprosium was substituted on the calcium sites [14-17].

156

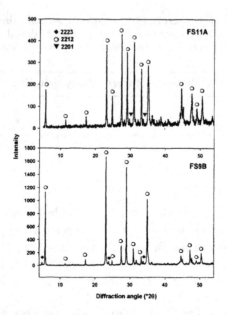

Figure 1. X-ray diffraction pattern of dysprosium-free (FS9B) and dysprosium for strontium substituted (FS11A) Bi-based 2223 materials, prepared from $Bi_{1.7}Pb_{0.3}Sr_{1-y}Dy_yCuO_x$ and $CaCuO_2$ via a two-powder process.

2.2. THE GLASSY ROUTE

In the "crystalline insert - glassy matrix method", we started from a glassy intermediate phase with two components ($Bi_{1.6}Pb_{0.3}Dy_{0.1}CuO_4 + 2$ CaO). The glass phase has been prepared by melting a homogeneous mixture of both components in an alumina crucible at 1050°C for a short time (30 min), and splat quenching the melt between two room-temperature copper blocks. The dysprosium doped Bi-based 2223 phase has then been prepared from a mixture of appropriate stoichiometric amounts of the ground glassy precursor and $SrCuO_2$. Next the mixture has been heat treated at 845°C during 6 days.

Figure 3 gives the X-ray diffraction pattern of the Dy-doped material (sample FS25B) as compared to the Dy-free similar compound (sample MS4B). Only the 2212 phase is clearly visible in the dysprosium doped material. The chemical composition has been analyzed by means of Energy Dispersive X-ray microanalysis. The observed Bi-based superconducting phase is essentially a dysprosium-doped 2212 material with a chemical formula close to $(Bi,Pb)_2Sr_2Ca_{0.9}Dy_{0.1}Cu_2O_z$. Secondary phases containing aluminium have been also noticed. It is thus concluded that the dysprosium substitution at the bismuth sites in the glass phase precursor inhibits the

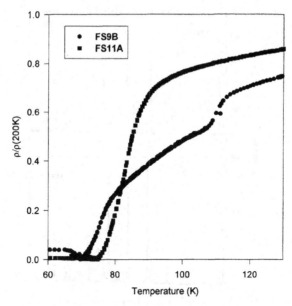

Figure 2. Electrical resistivity measurements as a function of temperature for dysprosium-free (FS9B) and dysprosium for strontium substituted (FS11A) Bi-based 2223
materials prepared by the two-powder process.

formation of the 2223 superconducting phase. Electrical resistivity measurements as a function of temperature are reported in Figure 4. Only one
superconducting transition is visible at 80K in the dysprosium doped material again, corresponding to the 84K or so transition expected for a pure
dysprosium doped BSCCO 2212 phase [18].

3. DISCUSSION

From the results presented in section 2, it can be deduced that the 2223
phase is "thermodynamically" unstable whatever the chemical route followed hereabove. It has to be noticed that the "non-formation" of the 2223
phase leads under the above experimental conditions to the growth of a dysprosium doped 2212 material. The trivalence of the rare-earth ion appears
to us to play a key role for the instability of the intended 2223 phase. The
instability reason can be searched for in difference(s) in the crystallographic
structures between the 2223 and 2212 phases (see Figure 5).

The basic difference between both structures lies in the presence of
central CuO_2 layers in square planar coordination for the 2223 phase only.
Substitution of dysprosium in these materials seems to be favored on the
calcium site whatever the selected substitution site. Such a substitution

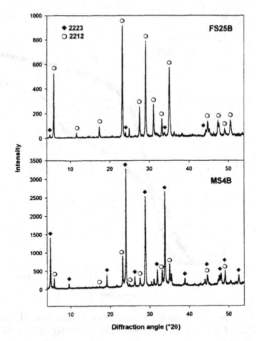

Figure 3. X-ray diffraction pattern of dysprosium-free (MS4B) and dysprosium for bismuth substituted (FS25B) Bi-based 2223 materials prepared from a mixture of appropriate amounts of the ground glassy $(Bi_{1.7-x}Pb_{0.3}Dy_xCuO_4 + 2\ CaO)$ precursor and $SrCuO_2$.

changes the overall charge distribution because of electron doping of the p-type CuO_2 layers. The hole density would thus decrease near the central CuO_2 layer. This would lead to a Cu-O bond length expansion, whence destabilizing the initial structure. Thus dysprosium doping would introduce antibonding electrons into the x^2-y^2 orbitals of the central CuO_2 layer. The intended structural transformation into a 2223 phase thus maximise the bond-length mismatch between the p-type superconductive layers, whence proving *ab absurdo* the difficulty of obtaining a dysprosium doped 2223 phase.

In fact, a similar type of behavior can be observed for strontium doping at the neodynium sites in the n-type $(Nd,Ce)_2CuO_4$ superconductor (characterized by the so-called tetragonal T structure, as shown in Figure 6) [19]. In this case, strontium doping on neodynium or cerium sites reduces the Cu-O bond-length of the fourfold CuO_2 sheets from 0.1975 nm to 0.1930 nm [20] leading to the formation of the well-known tetragonal T* structure (see Figure 6) by minimizing the bond-length mismatch between the constitutive layers. In both the fundamental reason is obviously the

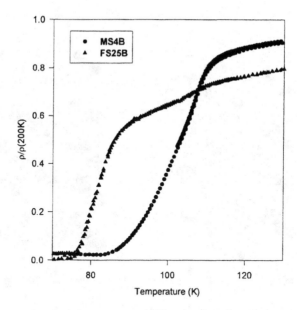

Figure 4. Electrical resistivity measurements as a function of temperature for dysprosium-free (MS4B) and dysprosium for bismuth substituted (FS25B) Bi-based 2223 materials prepared from a mixture of appropriate amounts of the ground glassy $(Bi_{1.7-x}Pb_{0.3}Dy_xCuO_4 + 2CaO)$ precursor and $SrCuO_2$.

minimization of the elastic energy [11,12].

The BSCCO case can be further understood by considering the data of Table II for the lattice parameters of both T and T* structures [19,20]. The c-axis lattice parameter is larger in the T* structure (1.250 vs 1.207 nm). This c-axis stretching can be related to the new crystallographic position following the motion of the apical oxygen away from the fivefold square-pyramidal CuO_2 layers as indicated by an arrow in Figure 6. This motion of the apical oxygen balances the electric change produced in the charge-carrier distribution by the bivalent strontium doping. In the case of the BSSCO related structures, the dysprosium substitution for calcium into the 2212 phase in contrast should induce an elongation of the planar Cu-O bond in the square-pyramidal CuO_2, an elongation which is balanced by the motion of the apical oxygen towards the central copper ion (see Figure 5), resulting in a c-axis reduction (the c-axis changes linearly with dysprosium content from 0.309 nm in Dy-free 2212 compound to 0.300 nm in fully "calcium by dysprosium" substituted compound) [21].

It is thus very important to control the charge distribution in the neighborhood of the superconductive layers so that the bond-length mismatch is minimized between the consecutive superconducting layers.

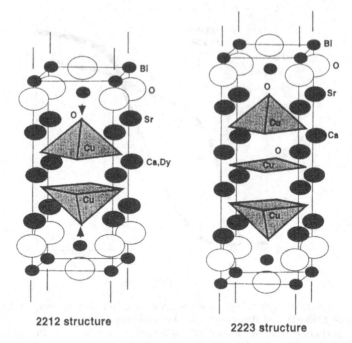

2212 structure

2223 structure

Figure 5. Crystal structures of the Bi-based 2212 and 2223 superconducting phases.

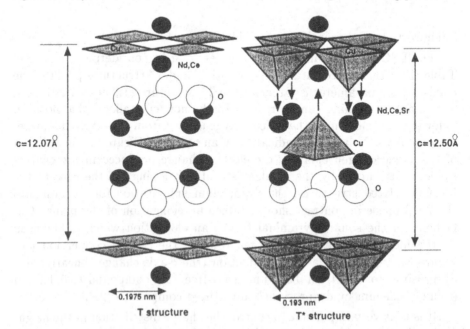

T structure

T* structure

Figure 6. Crystal structures of the $(Nd,Ce)2CuO_4$ (the so-called T structure) and the $(Nd,Ce,Sr)2CuO_4$ (the so-called T* structure) phases.

TABLE 2. Crystallographic lattice parameters of both $(Nd,Ce)_2CuO_4$ (T) and $(Nd,Ce,Sr)_2CuO_4$ (T*) tetragonal structures, as deduced from refs. [19] and [20].

Structure	Formula	a-axis (A))	c-axis (A)	space group
T	$(Nd,Ce)_2CuO_4$	3.95	12.07	I4/mmm
T*	$(Nd,Ce,Sr)_2CuO_4$	3.86	112.50	P4/nmm

4. CONCLUSIONS

It has been shown by X-ray diffraction analyses, Energy Dispersive X-ray microanalysis and electrical resistivity measurements as a function of temperature that intended doping of the Bi-based 2223 phase by dysprosium leads instead to the formation of the related Dy-doped 2212 phase. A structural instability resulting from electronically induced bond mismatch seems to be the fundamental cause. The instability depends on the ionic valence of the dopant and the selected crystallographic substitutional sites of the structure. The trivalent substitution induces a change in the charge carrier distribution near the superconductive CuO_2 layers which results in some structural deformation in terms of Cu-O bond-length variation. The variation of the Cu-O bond-length in the layers leads to a bond-length mismatch between the constitutive layers. This mismatch induces a stretch or shortening of the c-axis, depending on the substitution ion valency, in order to reduce the mismatch. Bond-length mismatch appears thus to play a very important role in determining which structure can be stabilized [22].

In the case of the 2223 phase, dysprosium substitution at the calcium, strontium or bismuth sites (which results in all cases in the formation of calcium partially substituted by dysprosium 2212 phase) induces an elongation of the Cu-O bond, and increases the elastic strain which can only be compensated by the "decomposition" of the architecture of the 2223 phase into a 2212 phase following apical oxygen motion in square-pyramidal CuO_2 sheets.

Acknowledgements

JF thanks the TMR program of the European Community for financial support.

References

[1] E. Farrell, B.S. Chandrasekhar, M.R. DeGuire, M.M. Fang, V.G. Kogan,

162

J.R. Clem and D.K. Finnemore, *Phys. Rev. B* **36** 4025 (1987)

[2] A. Luskinov, L.L. Miller, R.W. McCallum, S. Mitra, W.C. Lee and D.C. Johnson, *J. Appl. Phys.* **65** 3136 (1989)

[3] F. Chen, R.S. Markiewicz and B.C. Griessen, in *Superconductivity and Applications*, ed. by Hoi S. Kwo

[4] R. Cloots, A. Rulmont, C. Hannay, P.A. Godelaine, H.W. Vanderschueren, P. Regnier and M. Ausloos, *Appl. Phys. Lett.* **61** 2718 (1992)

[5] S. Stassen, R. Cloots, Ph. Vanderbemden, P.A. Godelaine, H. Bougrine, A. Rulmont and M. Ausloos, *J. Mater. Res.* **11** 1082 (1996)

[6] S. Stassen, A. Rulmont, T. Krekels, M. Ausloos and R. Cloots, *J. Appl. Cryst.* **29** 147 (1996)

[7] S. Stassen, A. Rulmont, Ph. Vanderbemden, A. Vanderschueren, Z. Gabelica, R. Cloots and M. Ausloos, *J. Appl. Phys.* **79** 553 (1996)

[8] Y.T. Huang, C.Y. Shei, W.N. Wang, C.K. Chiang and W.H. Lee, *Physica C* **169** 76 (1990)

[9] K. Yamada, K.I. Takada, S. Hosaya, T. Watanabe, Y. Endoh, N. Tomonaga, T. Suzuki, T. Ishigaki, T. Kamiyana and H. Asano, *J. Phys. Soc. Jpn.* **60** 2406 (1991)

[10] M. Wang, G. Xiong, X. Tang and Z. Hong, *Physica C* **210** 413 (1993)

[11] Z.-X. Cai and D.O. Welch, *Physica C* **231** 383 (1994)

[12] Z.-X. Cai, Y. Zhu and D.O. Welch, *Phys. Rev. B* **52** 13035 (1995)

[13] R. Cloots, H. Bougrine, M. Houssa, S. Stassen, L. d'Urzo, A. Rulmont and M. Ausloos, *Physica C* **231** 259 (1995)

[14] K. Nanda Kishore, M. Muralidhar, V. Hari Babu, O. Pena, M. Sergent and F. Bénière, *Physica C* **204** 299 (1993)

[15] K. Nanda Kishore, S. Satyavathi, M. Muralidhar, V. Hari Babu, O. Pena, M. Sergent and F. Bénière, *Phys. Stat. Sol. A* **143** 101 (1994)

[16] P.V.P.S.S. Sastry, J.V. Yakhmi and R.M. Iyer, *J. Mater. Chem.* **4** 647 (1994)

[17] L. Bonoldi, G.L. Calestani, M.G. Francesconi, G. Salsi, M. Sparpaglione and L. Zini, *Physica C* **241** 37 (1995)

[18] T.T.M. Palstra, B. Batlogg, L.F. Schneemeyer, R.B. Van Dover and J.J. Waszczak, *Phys. Rev. B* **38** 5102 (1988)

[19] Y. Tokura, H. Tagaki and S. Uchida, *Nature* **337** 345 (1989)

[20] F. Izumi, E. Takayama-Muromachi, A. Fujimori, T. Kamiyama, H. Asano, J. Akimitsu and H. Sawa, *Physica C* **158** 440 (1989)

[21] C. Kendziona, L. Forro, D. Mandrus, J. Hartge, P. Stephens, L. Mihaly, R. Reeder, D. Moecher, M. Rivers and S. Sutton, *Phys. Rev. B* **45** 13 025 (1992)

[22] J.B. Goodenough, in *Materials and Crystallographic aspects of High T_c Superconductivity*, ed. by E. Kaldis, NATO ASI Series vol. **263**, Kluwer Academic Publishers, Dordrecht-Boston-London, 1994, p. 175.

CRYSTAL GROWTH AND SOME PROPERTIES OF REMe₃ COMPOUNDS (RE = RARE EARTH, Me = Sn, Pb, Ga and In)

Z. KLETOWSKI
Institute of Low Temperature and Structure Research, Polish Academy of Science, 50-422 Wrocław, Okólna 2, Poland.

1. Introduction

The REMe₃ intermetallics crystallise with the fcc, AuCu₃ type of crystal structure. There are few series of compounds among them. The most numerous are REPb₃, REIn₃ and REGa₃. In the two last decades these systems have attracted a great deal of interest because of their salient feature such as valence fluctuation, superconductivity, magnetic moment formation, crystal field effects or multiaxial magnetic structures. The electronic structure of this family of compounds has been recently intensively studied by investigation of the de Haas-van Alphen effect and similarities in topology of their Fermi surface has been found [1].

Although magnetic properties of these systems are mainly determined by the simple RKKY interactions each compound in these series shows its unique magnetic properties due to different filling of the 4f shell or different crystal field influence. On the other hand, the same type of crystal structure and magnetic interactions in these systems along with similarities in their Fermi surface topology make the REMe₃ intermetallics suitable for investigation of the influence of systematically changed RE ion on different electron transport properties. Such investigations of the electrical resistivity and the thermopower are presented.

2. Single Crystal Growth

The single crystals of REMe₃ intermetallics used in our experiments were obtained by the method of growth from the molten solution prepared of non-stoichiometric amounts of substrates. The method has a few advantages: (i) It reduces the cost of the used apparatus. (ii) It minimises the problem off stoichiometry caused by the oxidation or evaporation of the RE element since the single crystal controls stoichiometry itself. (iii) The crystallisation process proceeds in a wide temperature range even if a significant excess of the Me element is used since the REMe₃ stoichiometry is the richest in the Me element. An excess of the Me element essentially lowers the melting temperature of the solution as well as the reactivity and the vapour pressure of the RE element. Furthermore

163

R. Cloots et al. (eds.), Supermaterials, 163–172.

it limits demands on technical parameters of the heating coil in the furnace as well as on refractoriness of the used crucibles and ampoules. (iv) The Me excess dissolves impurities originating mainly from the RE element. This results in higher purity of the $REMe_3$ crystals than could be anticipated from purity of the RE component used. In our experiment the off stoichiometry ranged from 3% to 15% of the Me excess. The best results we obtained concerning size and shape of crystals for the 5-6% excess. The maximal operating temperature was 1080°C, what enabled us to use corundum crucibles, quartz glass tubes, and a common resistant wire to wound heating coil. A cross section of the furnace is shown in Fig. 1a. This furnace was constructed in the way to protect its high thermal inertia. An additional heat resistant steel container, placed inside the furnace, was used to improve the temperature stability at the place of crystal growth.

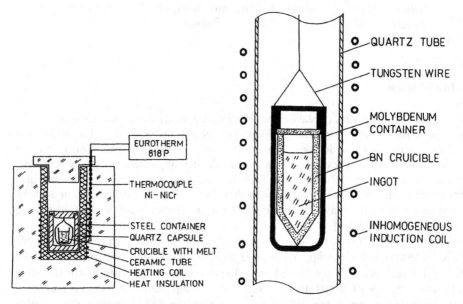

Figure 1a. Cross section of the furnace used to growth the $REMe_3$ compounds.

Figure 1b. Experimental arrangement for growth of the $CeCu_2Si_2$ single crystals.

The furnace temperature was controlled by the Eurotherm 818P connected with the Ni-NiCr thermocouple placed very close to the heating coil. An extra thermocouple, placed inside the steel container, was used for a precise measurement of the temperature profile while the furnace was thermally tested. Temperature oscillations measured inside the steel container were lower than 0.05°C and not dependent on the cooling rate.

A well outgassed corundum crucible containing the RE and the Me substrates was sealed in a quartz ampoule under 150mmHg argon pressure and placed inside the steel container of the furnace.

At the commencement of the procedure the ampoule containing a chosen RE - Me composition was heated to temperature of about 10°C above the liquidus - solidus curve T_{ls}. Then, temperature was lowered to about 10°C below T_{ls} and the procedure proposed by Scheel and Elwell [2] was applied in order to minimise the number of crystallites formed on nucleation. Within a small temperature interval, the temperature

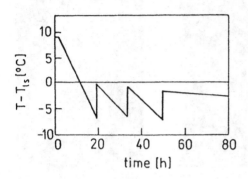

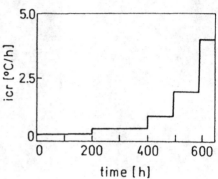

Figure 2. Temperature program to minimise the number of crystallites formed on nucle--ation.

Figure 3. Time dependence of the cooling rate.

was repeatedly decreased at slow rate, and than quickly increased again. After 3-4 such treatments the main cooling program was started with the initial cooling rate, icr. The cooling rate was progressively increased up to 5°C/h until the ampoule reach temperature of the REMe₃ - Me eutectic, then the system was switched off. An example of this cooling procedure is shown in Fig. 2 and 3. After opening the quartz ampoule the REMe₃ crystals were extracted from the ingot by alloying the Me - excess matrix with gallium in the case of the indium matrix, or with an Wood-like alloy (Bi:Pb:Sn:Cd = 4:2:1:1) in the case of a tin or lead matrix. To avoid an increase of the melting temperature of the Wood-like alloy because of composition change, the alloy was prepared with a deficiency of Sn and Pb elements. Thus it was possible to melt the Me matrix at temperatures below 100°C. Surfaces of the extracted single crystals remained covered with a thin layer of the alloy. This layer was finally removed by either brushing out the crystals, kept on a heated support in an inert atmosphere, or by washing them in mercury and final desorption of the mercury in high vacuum.

Another method of removing the Wood-like alloy was developed by Y. Onuki. The warmed up crystals are rotated in an centrifuge with the velocity of about 85rad/sec [3]. Such rotation removes much better the alloy from the crystal surface than the brushing. Crystals cleaned in such a way are pure enough for the dHvA experiment. However, for electron transport investigations, an additional chemical treatment is necessary to remove completely the alloy from their surface.

Our experiments on the single crystal growth of the REMe₃ [4] compounds began with the Me=Sn and for relatively high icr=1.3°C/h and vertical temperature gradient, vtg=1.5°C/cm. Under these conditions all RESn₃ crystals grew vertically along the crucible in a dendrite-like form.

166

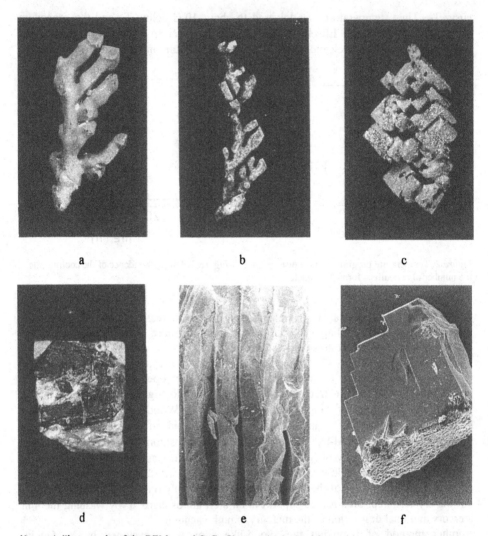

a   b   c

d   e   f

Figure 4. Photographs of the $REMe_3$ and $CeCu_2Si_2$ crystals obtained for various values of initial cooling rate, icr and vertical temperature gradient, vtg. Fig. 4a and 4b show dendrite-like form of $PrSn_3$ and $NdSn_3$ crystals, respectively obtained for high values of the icr and vtg. Fig. 4c shows platelets of the $CeSn_3$ crystals. Fig. 4d shows cuboid form of the $GdIn_3$ crystal obtained for low values of the icr and vtg. Fig. 4e and 4f show plate-like crystals of $CeCu_2Si_2$ grown by slow cooling of the melt. Thickness of the plates shown on Fig. 4e is 0.7mm. The plate shown in Fig.4f is 1mm thick.

After reducing the icr and the vtg to 0.4°C/h and 0.3°C/cm, respectively a change of form of growing crystals across the series was observed.

The LaSn₃ crystals grew as cuboids with the largest dimension up to 4mm. The CeSn₃ crystals grew in the form of platelets of a non-regular hexahedron shape, with the longest diagonal, 6mm, along the [110] direction. Generally, the heavier the RE element was, the smaller the crystals of the RESn₃ compounds were. They grew along the crucible and the [100] and [110] axes were typical growth directions.

For the REIn₃ compounds the icr and vtg parameters were decreased to 0.2°C/h and 0.1°C/cm, respectively. Under such conditions the single crystals grew in the cuboid form only. They had well formed (100) type planes and the largest dimension up to 8mm. Unfortunately, the largest crystals of REIn₃ usually contained a few indium inclusions with size of a few tenth of millimetre.

For growing the REPb₃ compounds similar set of icr and vtg parameters was used as in the case of growth of the REIn₃ ones. The REPb₃ grew in the cuboidal form, however their size was about two times smaller as that for the REIn₃. The [100] direction was typical growth in this case.

For commonly used purity of the components equal 3N for the RE element and 5N for the Me element the residual resistivity ratio rrr=ρ_{300}/ρ_0 , ρ_0=residual resistivity, of the obtained crystals was found in the range between 20 and 150. However most crystals had their rrr between 45 and 85. Such rrr value and a strain balanced crystal lattice obtained during the very slow cooling process enabled us to investigate easily all electron transport properties of these crystals and also their electronic structure by means of the de Haas van Alphen effect.

The macrophotographs from 4a to 4d show a change of crystals shape obtained for various icr and vtg parameters.

The single crystal growth from solution is especially fruitful in a case of incongruently melted compounds when common methods of the growth as the Bridgeman or Czochralski ones are inefficient. An example of the application of this method of crystallisation for such a purpose is the growth of the well known CeCu₂Si₂ heavy fermion compound that melts incongruently at 1545°C.

We grew single crystals of this compound from solution in a Bridgeman shaped crucible. Polycrystalline samples of the compound were prepared from appropriate quantities of the constituents in an arc furnace under an argon atmosphere. These samples were used as the starting material to load the boron nitride, BN crucible (BN crucible with an inner diameter of 7mm). The crucible was then placed in a molybdenum container with a wall thickness of 2mm. The container, which was next situated in an inhomogenous induction coil, was used as the heat source. Fig. 1b shows the details of this arrangement. The loaded crucible was heated under argon atmosphere to about 40°C above the melting point of the solution. This point, which was estimated to be about 1470°C, was indicated by slight cooper evaporation. After the induction furnace and the crucible had reached thermal equilibrium, the furnace power was reduced continuously so the molybdenum container, which was hanging motionless, was cooled down. The initial cooling rate was in this case 8°C/h. The temperature was reduced in such a manner to 1250°C and then it was reduced to room temperature much quicker with the rate of

$100°C/h$. Due to a special profile of the induction coil different values of the vertical temperature gradient were possible to obtain.

In result, plate-like single crystals of the $CeCu_2Si_2$ were grown. They grew along the crucible parallel to each other and had a mirror-like (110) basic plane. The c axis was always perpendicular to the platelets. Size of the crystals was about 5mm in the basic plane and up to 1mm in the c direction. The best results were achieved with the vertical temperature gradient of $15°C/cm$. However, this vtg refers to the outer container wall temperature and was measured using an optical pyrometer.

The scanning electron micrographs 4e and 4f show typical shape and size of the $CeCu_2Si_2$ single crystals grown in such manner.

3. Resistivity and Thermopower Properties

We present a few results concerning resistivity and thermopower (TEP) obtained for the heavy $REIn_3$ compounds. Some regularities observed here are consequence of gradually filling up the well localised 4 f shell. Since this group of compounds display close similarities in the Fermi surface and phonon properties therefore they are good candidates for systematic study of resistivity and thermopower.

A separation of the magnetic part of the TEP is simpler if we restrict our attention to the paramagnetic phase and temperatures higher than the Debye temperature, ca.150K, in these compounds. The magnetic part of the resistivity at these temperatures is temperature independent due to an overall population of the crystal field levels, see Fig. 8. As far as the TEP is concerned, the diffusion term is dominant in these temperatures. In such conditions, assuming validity of the Matthiessen rule, one can reasonably apply the Nordheim - Gorter rule in order to separate the magnetic term of the TEP. After Gratz et al. [5] thermopower for the nonmagnetic compound, S^{nm} as well as for the magnetic one, S^m is given as

$$S^{nm} = (\rho_o/\rho)S_o + (\rho_{ph}/\rho)S_{ph} \tag{1}$$

$$S^m = (\rho_o/\rho)S_o + (\rho_{ph}/\rho)S_{ph} + (\rho_m/\rho)S_{SPD} \tag{2}$$

The ρ_o, ρ_{ph} and ρ_m denote resistivity contributions associated with scattering on impurities, phonons and magnetic moments respectively. The S_o, S_{ph} and S_{SPD} represent analogous terms in the TEP. When assuming that the phonon and impurity contributions are respectively equal for the nonmagnetic standard and magnetic $REIn_3$ compound we obtain the spin dependent part of the TEP, S_{SPD} (the magnetic part) as

$$S_{SPD} = [S^m - S^{nm}] \, \rho/\rho_m \tag{3}$$

Here ρ is the total resistivity of the $REIn_3$ compound, ρ_m is the magnetic part of the resistivity which at the considered temperatures is a constant equal to the spin disorder resistivity $\rho_{m\infty}$. In the case of the heavy $REIn_3$ compounds such approach is, in a first approximation, reasonable because close similarities between all S(T) curves are observed in the temperatures higher than the Debye temperature, see Fig. 5, the resistivity dependencies $\rho(T)$ show the same slope at these temperatures, see Fig. 6, and finally the investigated samples are of equal purity.

The S_{SPD} contributions in the heavy $REIn_3$ compounds calculated according to

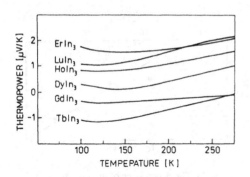

Figure 5. Temperature dependence of the thermopower of the heavy REIn₃ compounds.

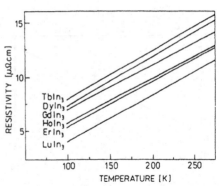

Figure 6. Temperature dependence of the resisti- -vity of the heavy REIn₃ compounds.

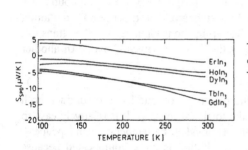

Figure 7. Temperature dependence of the mag- -netic term of the thermopower, S_{SPD} for the heavy REIn₃ compounds.

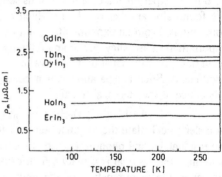

Figure 8. Temperature dependence of the mag- -netic term of the resistivity for the heavy REIn₃ compounds.

the formula (3) are plotted in Fig. 7. The results show that the S_{SPD} term is linearly dependent on temperature and the slope of this dependence is the same for the all investigated compounds within the experimental error. If one plots the values of the S_{SPD} at the highest applied temperature (T=295K) against the de Gennes factor one gets the straight line, see Fig. 9. Usually such a linear dependence is observed for the spin disorder resistivity as an exemplification of the RKKY type s-f interactions. In the heavy REIn₃ compounds we also observe such behaviour of the spin disorder resistivity, confirming the RKKY type of interactions as well as similarities in the electronic structure of these compounds, see Fig. 10.

The linear dependence of the magnetic part of the TEP with the de Gennes factor found experimentally was not reported earlier. This first experimental data may suggest that TEP behaves in the same way as the spin disorder resistivity not only in the REIn₃

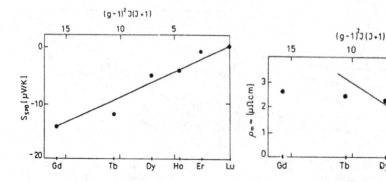

Figure 9. The dependence of the magnetic term of the thermopower on the de Gennes factor.

Figure 10. The dependence of the spin dis-order resistivity on the de Gennes factor.

compounds, but also in other f electron systems. However to verify this conclusion one need more experimental data for various series of f electron systems.

We found also another regularity which concerns the nonmagnetic $LaMe_3$ and $LuMe_3$ compounds. From an experimental point of view the properties of these compounds are important because they are commonly used as standards for description of phonon scattering in the separation of various scattering contributions to the transport coefficients. Such a separation which concerns the TEP was given above. An another one concerns the electrical resistivity.

$$\rho(T) = \rho_o + \rho_{ph}(T) + \rho_m(T) \tag{4}$$

In order to calculate the magnetic term of the resistivity, ρ_m one have to subtract both the residual, ρ_o and phonon, ρ_{ph} terms from the total resistivity, ρ. It is relatively easy to determine the residual resistivity which can be assumed to be temperature independent. However, the determination of the phonon contribution is more complicated because the temperature dependence of ρ_{ph} can be specific for a particular compound. For this purpose either the resistivity of an nonmagnetic homologous compound of the investigated one or the Bloch - Grüneissen, B-G function is commonly used. By examining as an example the $PrIn_3$ we present how big discrepancies occur if various phonon standards are used to determine experimentally the magnetic part of the resistivity ρ_m. The $PrIn_3$ was chosen because it is a singlet ground state system which does not order magnetically and the magnetic part of the resistivity is mainly caused by the influence of the crystal field effect on the scattering process. For such a system the $\rho_m(T)$ dependence can be easily calculated theoretically in the whole temperature range when the crystal field level scheme is known. Fig. 11 shows the result of the final separation procedure. The magnetic contribution to the resistivity is given here as a part of the spin disorder resistivity $\rho_{m\infty}$. The curves A, C and D represent the magnetic contribution to the resistivity which were determined for three different phonon standards. If the resistivity of $LuIn_3$ was used as the phonon standard we obtain curve A, if the resistivity of the $LaIn_3$ - curve C and finally, if the B-G function was taken as the phonon standard - curve D, respectively. Curve C represents theoretical predictions calculated after N.H. Andersen *et al.* [6].

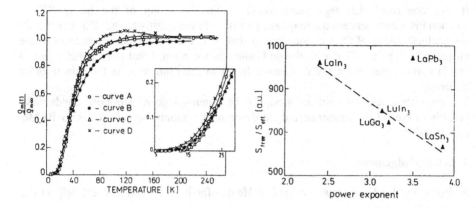

Figure 11. Comparison of the temperature depend-
-encies of the magnetic part of the resistivity ob-
-tained experimentally for various phonon standards
in PrIn₃. See comment in the text.

Figure 12. Correlation between the free electron
to real electron Fermi surface ratio and the
power exponent in the generalised BG function.

It is well seen that different results were obtained for the different reference materials.
That was the reason why we began to investigate the nonmagnetic REMe₃ compounds
more intensively. Our goal was to improve the existing methods of the separation.
Any theory predicting physical relations will be useful as a guide for establishing
experimentally approved extrapolation function. We found that such a function is
the generalised B-G function where the power exponent instead of 5 is allowed to
assume noninteger values. Although there are no physical grounds for this assumption,
such a fit function describes much better the phonon term than commonly used reference
standards [7]. Analysing data on the resistivity of five nonmagnetic REMe₃ compounds
we found that such noninteger power exponent is linearly dependent on the free electron
to real electron Fermi surface ratio of the compound under investigation.
The dependence is shown in Fig. 12. The free electron to real electron Fermi surface
ratio was determined according to formulas given by Ziman [8]. The figure 12 strongly
suggest that it might be possible to give a prescription how to determine the power
exponent of the $\rho_{ph}(T)$ dependence if one knows the value of its high temperature
derivative. The latter with the Debye temperature incorporate the free electron to real
electron Fermi surface ratio and finally from the Fig. 12 one can read the power
exponent in the generalised B-G function for a certain REMe₃ compound. Now we
search for similar regularities in other f electron systems.

4. Conclusions

Single crystals of the REMe₃ intermetallics were obtained in a relatively simple method
of the crystallisation from molten solution. An essential influence of both the vertical
temperature gradient and the initial cooling rate on the crystallisation process was
established.

It was confirmed that regularities found within the group òf the heavy REMe₃ compounds which concern the magnetic part of both the resistivity and TEP come from the gradually filling of the well localised *4f* shell. The observed linear dependence of the magnetic term of the TEP with the de Gennes factor suggest that the magnetic part of the TEP is dependent on the de Gennes factor in the same way as the spin disorder resistivity.

The correlation found within the group of the nonmagnetic REMe₃ compounds make possible to predict the temperature dependence of the phonon term in the other REMe₃ compounds.

5. Acknowledgement

Author is indebted very much to prof. Z Henkie for his kind and efficient help in this work.

6. References

1. Onuki, Y. and Hasegawa, A. (1995) Fermi surfaces of intermetallic compounds, in K.A. Gschneider, Jr and L. Eyring (eds.), *Handbook on the Physics and Chemistry*, vol. 20, Elsevier Science B.V. pp. 1-103.
2. Sheel, H.J. and Elwell, D. (1972) Stable growth rates and temperature programming in flux growth, *Journal Crystal Growth* 12, 153-161.
3. Onuki, Y. *private communication*.
4. Kletowski, Z., Iliev, N., Henkie, Z. and Staliński, B. (1985) Single crystal growth of (Rare Earth)Me₃ compounds, *Journal of the Less-Common Metals* 110, 325-238.
5. Fournier, J.M. and Gratz, E. (1993) Transport properties of rare earth and actinide intermetallics, in K.A. Gschneider, Jr., J. Eiring, G.H. Lander and G.R. Chopin (eds) *Handbook on the Physics and Chemistry of Rare Earths*, vol 17, Elsevier Science B.V.
6. Hessel Andersen, N., Gregers-Hansen, P.E., Holm, E. and Smith, H. (1974) Temperature-dependent spin-disorder resistivity in a van Vleck paramagnet, *Physical Review Letters* 32, 1321-1324.
7. Kletowski, Z., Fabrowski, P., Sławiński, P. and Henkie, Z. (1997) Resistance of some REMe₃ compounds, RE = La and Lu, Me = Sn, Pb, In and Ga, *Journal of Magnetism and Magnetic Materials*, 166, 361-364.
8. Ziman, J.M. (1962) *Electrons and Phonons*, Oxford University Press, Oxford, p. 374.

PARTICULAR ASPECTS IN NiMnSb SEMI-MAGNETIC ALLOY RELATED TO THE GROWTH CONDITIONS

Review of some experimental results

C. GRIGORESCU[1]*, C. LOGOFATU[2], S.A. MANEA[3],
M.F. LAZARESCU[3], N. STANICA[4], R. NOTONIER[5],
J-C. MARTY[5], A. GARNIER[5], L. TORTET[5],
O. MONNEREAU[5], G. VACQUIER[5], A.C. ROWE[6],
R.A. STRADLING[6]

[1] *National Institute for Optoelectronics, Bucharest, Romania*
[2] *National Institute for Plasma, Laser and Radiation Physics, Bucharest, Romania*
[3] *National Institute for Materials Physics, Bucharest, Romania*
[4] *Institute of Chemistry Physics "G. Murgulescu", Bucharest, Romania*
[5] *Universite de Provence, Marseille, France*
[6] *Imperial College of Science, Technology and Medicine, London, United Kingdom*

1. Introduction

Ternary Heusler alloys of the type XMnSb (X=Ni, Pt, Cu, Co, Au, Fe) [1-3]exhibit very promising magnetic, magneto-optic and electric properties for magnetic device and data storage applications.

In the early 50's it was shown that from all combinations mentioned above the NiMnSb alloy displays the strongest ferromagnetic effect when the stoichiometric composition 1:1:1 is achieved [1]. The high Curie temperature, ranging from 693K to 763K - according to the composition and, correspondingly, to the crystallographic structure - makes the material particularly interesting in both thin layers and bulk form. Besides, the virtual 100% spin polarisation would encourage spin-device applications.

Our present work is concerned with particular aspects of bulk NiMnSb related to the growth conditions, which seem less touched in previously published papers.

We stress on the experimental aspect of this work, so that any detail that might not match the "classical" knowledge on the compound should be taken as an reproducible experimental fact, though sometimes a clear explanation is not yet available. This arises mainly from the novelty of the results that opens the investigation field further on.

As it is known [1] oxidation during synthesis and growth should be avoided as much as possible, and that is why usually aluminum oxide crucibles and inert atmosphere have been used.

We made an attempt to study the effects of quartz and vacuum on the direct synthesis of the alloy in comparison with the "classical" conditions. Various thermal regimes were experienced to find out the best solution in growing homogeneous polycrystalline NiMnSb with 5.00 to 17.00 mm diameters with the aim of preparing suitable targets for laser deposition techniques and, not in the least, to basicaly investigate the behaviour of the elements when building up the compound.

The alloys were obtained using a MSR-2 (Cambridge Ltd.) equipment and respectively a home-made plasma furnace whose description is detailed in the reference[4].

We checked the structure and the composition of the samples by XRD and EDAX respectively. Segregation of antimony in some of the samples has been put in evidence by SEM and confirmed by Raman spectroscopy. We do not enter the details of the lattice dynamics in this work. A forthcoming paper will be concerned with these aspects, since the lattice dynamics subject has not been touched yet to our knowledge and it makes itself a particular problem.

The variation of the electric resistance of our alloys was measured over the temperature range 20K to 300K, and magnetic measurements were performed at room temperature. The effect of

R. Cloots et al. (eds.), Supermaterials, 173–182.
© *2000 Kluwer Academic Publishers. Printed in the Netherlands.*

chromium up to 20% in the NiMnSb has also been investigated, mainly in respect of the electric and magnetic properties of the "quaternary" alloy, showing that it preserves the semimetallic/ferromagnetic aspects.

2. Experimental

2.1 PREPARATION OF THE ALLOYS

To achieve our purpose we used high purity Ni, Mn, Sb in the stoechiometric proportions 1:1:1 for the NiMnSb alloy and Ni-Cr (20% for Cr); Mn ans Sb in the $Ni_{1-x}Cr_x$ MnSb batches.
The attempts to grow the alloys were made:

i)   using the classical alumina crucibles, conical in shape (3/5 radii ratio); the melting/crystallisation process developed in nitrogen at the atmospheric pressure;

ii)   using double walled quartz ampoules, cylindrical in shape (6 mm inner diameter); all processes happened under vacuum.

The reason to change the material and the shape of the crucible is to find out the influence of the thermal gradient on the segregation phenomena in this system, having in view the so different thermal properties of alumina and quartz, such as thermal conductivity and expansion coefficient.

As specified in the previous section, we also made use of two different equipments in our experiments. A more detailed description of the runs is given below.

The accurately weighted ammounts of Ni, Mn, and Sb were encapsulated in the duble walled quartz ampoules and sealed under vacuum. Three identical batches were prepared this way with the aim to run the melting/crystallisation processes with both the MSR-2 (RF heating) and with the plasma furnace.

For the batch containing the chromium alloy we used $Ni_{0.80}Cr_{0.20}$ sheet.

In what follows the samples we investigated are denoted as in the table 1:

TABLE 1. Description of the samples in connection with the material of the crucible and the type of furnace used

Sample	Assumed composition	Diameter [mm]	Crucible	Heating system
NMS1b	NiMnSb	16	alumina	RF (MSR-2)
NMS1t	NiMnSb	17	alumina	RF (MSR-2)
NCMS1	$Ni_{0.80}Cr_{0.20}MnSb$	6	quartz	RF (MSR-2)
NCMS2	$Ni_{0.80}Cr_{0.20}MnSb$	6	quartz	plasma

It is to emphasise that from all the crystals we made the above selection represents the most eloquent one to illustrate the subject of this paper.

The reason to include Cr in these experiments is explained on the basis of some questions arisen along with our work:

i)   why was this element not involved in formerly prepared XMnSb alloys, having in view its electronic structure?

ii)   which way will follow the structural, electric and magnetic properties of an alloy containing additional chromium in comparison with those of the true NiMnSb?

iii)   how will behave the quaternary alloy in respect of growth conditions in comparison with the ternary NiMnSb?

Certainly the reader might ask "why $Cr_{0.20}$ and not simply Cr instead of Ni?"

The answer is that for a start with the new element we did not want to go too far away from the basic NiMnSb alloy and in the same time a quaternary alloy makes a particularly interesting experience for a grower.

To describe the thermal regimes followed with the batches it is appropriate to look at the graphs in *Figure 1*.

The curve numbered "1" was used for the batch sampled as NCMS1 and that one numbered "3" - for the batch denoted by NCMS2 (see the table 1.).
Both regimes show a similar behaviour in the cooling stage, i.e. a steep decrease of the temperature which was aimed to enhance the chance to preserve the composition held by the alloys at the melting point. Whether it happened or not we shall see in the section devoted to the characterisation of the samples.
The short plateau shown by the regime "1" was necessary to enable natural convection to act more intensively, while in number "3" it does not show itself up - the setup holding a plasma heater takes usually advantage of the effects of this phenomenon happening very fast due to the geometry of the furnace which provides a particular heat exchange.

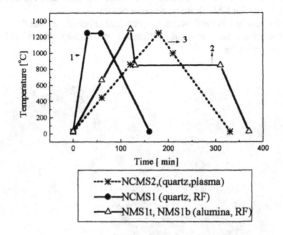

Figure 1. The thermal regimes used to prepare the alloys.

The curve "2" is more special in shape in comparison with the previous ones. Heating up to 1300°C of the batch took place more slowly and a larger plateau was set at 850°C, after decreasing the temperature suddenly to the mentioned value. We remind the reader that the ammounts of components were much larger than in the previous cases and the crucible was made of alumina. These details provide a few information on the heat transfer and the thermal equilibrium problems arisen from the material of the crucible and from the alloy mass. We also should underline that the temperature regime number "2" follows in part that one described in the reference [1], but the synthesis and the crystallisation of the final product happened in one run only in our case.

2.2 CHARACTERISATION OF THE SAMPLES

2.2.1 *Structure investigation*
The XRD spectra of the samples in Table 1, taken with the CuKα1 radiation of a Siemens - D5000 diffractometre, show all the major features attributed to NiMnSb [2] and this happens for NCMS too. However, the "cleanest" one - phase NiMnSb spectrum belongs to NMS1b, while for NMS1t we rather notice the Ni$_2$MnSb than the 1:1:1 composition, though one - phase too.
As for the NCMS1 and NCMS2 - they preserve the cubic structure of the intended compound.
The cubic structure has a great importance for the magnetic properties of these XMnSb alloys, but the right composition and consequently the occupation of the sites in the three cubic sublatices, assumed for

the NiMnSb like pseudo-ternary alloys has an even stronger influence on their magnetisation and Curie temperature respectively [1,3]. On this basis we looked by EDAX analysis at our samples in connection with SEM images and micro-Raman spectroscopy and further on we measured their magnetisation at room temperature.

EDAX, SEM and Raman investigations. The EDAX analyses following the SEM imaging of the samples were performed using a XL 30 ESEM LaB_6 (Philips Optique Electronique - France) equipment.

As one observes in *Figure 2, (a, b and c,* corresponding respectively to NMS1b, NCMS1and NCMS2), there is a major difference between the images.

The aspect of the surface in NCMS1 shows a lamellar structure, whose EDAX analyses will be further detailed. We also noticed a quite strong segregation effect of antimony in this sample. This was not observed at all either in NMS1b (*Figure 2 - a*), or in NMS1t. We do not show here the image of the latest one because of its high similarity with NMS1b from this point of view.

As for NCMS2, the lamellar structure is much weaker represented (*Figure 2-c*) and the EDAX analyses performed on this sample show a quite interesting difference as compared to NCMS1, though both samples were grown in quartz ampoules.

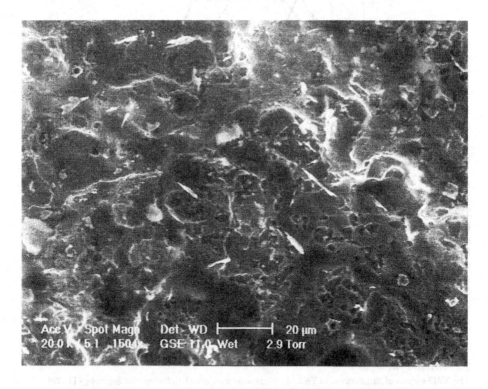

Figure 2 a) - SEM image taken from NMS1b (polycrystalline NiMnSb grown in alumina crucible)

177

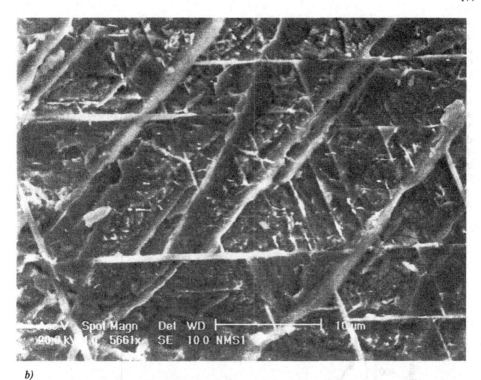

b)

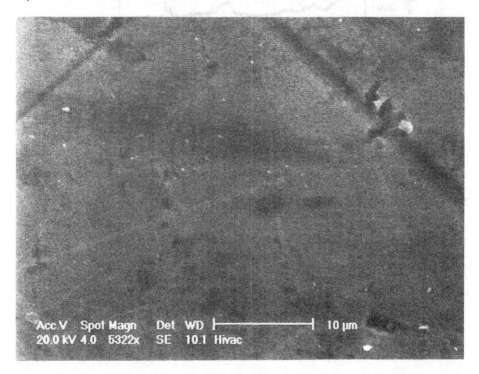

c)

Figure 2 b), c): SEM images of NCMS1 (b) and NCMS2 (c) which were grown in quartz ampoules.

Since the micro-Raman spectroscopy makes a very fine tool in respect of surface characterisation, we wanted to make sure that in the regions where no Sb is obvious in NMS1 it happened actually NO antimony segregation.

On this purpose the Raman spectra were taken with a RENISHAW 2000 ramascope (Renishaw Ltd., The United Kingdom).

We used in the experiments the backscattering configuration and an Ar^+ laser source at 514.5 nm. As seen in the Figure 3, the surface of NMS1b still exhibits a Sb segregation effect (the peaks at about 116 cm^{-1} and respectively 150 cm^{-1} corresponding to crystalline antimony), but much weaker than in the case of NCMS1. However, looking at the data in the Table 1, one could expect the oposite due to the large difference between the diametres of the samples.

We believe the weaker segregation of antimony arises from both the crystallisation regime we used to make the NMS1 and the material of the crucible, i.e. alumina, whose thermal properties are more favourable to the homogeneity of the thermal field in the system. It is not the point of this work to discuss in more detail the Raman spectra, but the improved crystalline quality of NMS1b in comparison with that one of NCMS1 has to be underlined starting from this figure.

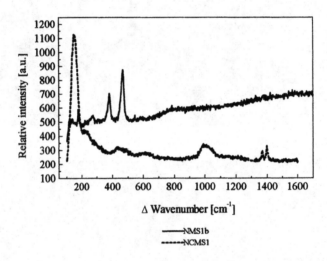

Figure 3. The Raman spectra of NMS1b and NCMS1 showing the presence of Sb.

The samples were further investigated with regards to the homogeneity of the respective batch and of the radial composition.

The EDAX measurements performed on NMS1(top and rear slices are denoted by "t" and "b" respectively) are shown in Figure 4 - a ("t") and b ("b") and for NCMS1 and NCMS2 in Figure 5, *a* and *b*. Remembering that synthesis and crystallisation of NMS1occured with one run only, it is obvious from these representations that the top slice is a different phase from the rear slice, and this is reflected by the magnetic measurements presented in a following section of this work.

At the moment however we focus on what arises from the Figure 4:

i)   *the phases are different with respect to the hight of the ingot (i.e. crucible)*

ii)   *the radial distribution of the composition is mostly satisfactory in the rear slice, which can be attributed to the true NiMnSb phase (see also reference[1]).*

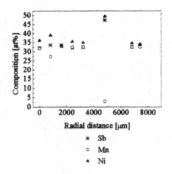
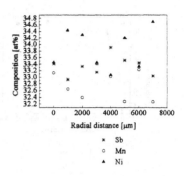

a)   b)

Figure 4. a) NMS1t; b) NMS1b
The radial distribution of the components in NMS1t and NMS1b
as it resulted from the EDAX analysis of the samples.

The difference between the composition of these two slices might be put on the conical shape of the alumina crucible that modifies the axial thermal gradient in the crystal growth equipment according to the 3/5 ratio of the bottom/top diameters respectively. Remarkable, the lower part of the melt had been better mixed and the radial segregation of the elements - particularly of antimony - happened at a much more reduced rate than in the upper region.

From the experimental results we actually believe that there had been formed two axial convection vortexes in the melt, acting constantly most likely due to the fixed position of the crucible in the growth setup.

On the other hand, the radial convection seems less intense in the smaller diameter region than in the larger one (though the difference between them is 1.0 mm only) but this does not look surprisingly: the distribution of the thermal field is of better homogeneity in this part of the system accounting for the heat transfer between two "successive" layers of the melt, while in the top zone the heat transfer occurs between the surface of the melt and the nitrogen in the chamber thus allowing the radial distribution of the thermal field to correspondingly change, but not in a favourable to our purpose way.

Coming back to the cylindrical samples (NCMS1 and NCMS2) grown in quartz under vacuum, we must say that the EDAX analyses along the radius gives much stranger results as seen in Figure 5-a) and b).

It is to notice the higher homogeneity in respect of components distribution of NCMS2 as compared to that of NCMS1. This is in our opinion a consequence of the thermal regime rather than of the equipment itself.

The very peculiar surface structure of NCMS1, which is shown in the *Figure 2- b)*, was quantitively analysed and surprisingly presents a particular behaviour of this alloy, as it results from the Table 2.

TABLE 2. The composition of the two prominent regions exhibited by NCMS1

NCMS1	Ni [at%]	Cr [at%]	Mn [at%]	Sb [at%]
light zone	7.7	21..35	28.24	39.23
(edge)	7.99	22.15	29.28	40.58
dark zone	28.17	3.68	33.05	33.21
(inside)	28.73	3.75	33.70	33.82

The analysis was repeated in several points of the cells confirming the examples in the table above.

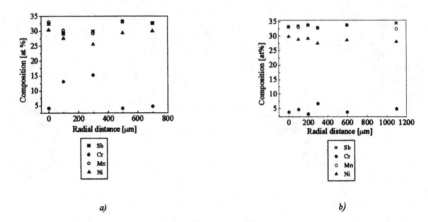

Figure 5. Radial distribution of the elements in NCMS1 (a) and NCMS2 (b)

Apart from the presence of silicon up to 3% (at.) found in both NCMS1 and NCMS2, whose source we believe at the moment to be rather traces of the polishing matters than the interaction with the quartz wall, it is to be noticed the interchange, roughly speaking, between chromium and nickel, outside and inside the observed structures in NCMS1. This looks like during the melting-crystallisation process Cr would segregate preferentially on the "edges" of the cells. The segregation effect is not observed in NCMS2, as it is clearly seen from the *Figures 2* -c) and *5-b)* respectively, and we relate this fact to the thermal regime too.

At a better look, it seems that the quaternary alloy can actually be represented as $Ni_{1-x}Cr_xMnSb$ in NCMS2, having in view the quite uniform distribution of the elements in this bulk sample.

In support of our former supposition with respect to the Si traces it is to be mentioned that no traces of oxigen were found from the analyses.

Electric and magnetic measurements. We measured the variation of the electric resistance of the samples over the temperature range 20K to 300K using the four probes method. The samples, semicircular in shape, were neither identical with respect to diameter nor to thickness (NMS1b is much larger than the other ones), but the graphs in Figure 6 look very similar apart from the values: a typical semimetallic behaviour is exhibited by all samples. At this stage the comparison is qualitative only.

The magnetic measurements were performed using the AC permeametre method, that is a typical electromagnetic induction means [5] and works correctly when the magnetic relaxation time is shorter than the AC period. The measured signal follows the proportionality given below:

$$\sigma_s = K_{instr}\varepsilon_{ef} \tag{1}$$

where σ_s is the magnetisation of the sample, ε_{ef} is the effective value of the electromotive force (which is the directly measured quantity with this method) and K_{instr} is the constantant of the permeametre. In our case, $K_{instr} = 0.13$.

The measurements were carried out at room temperature, but no doubt future experiments are needed over large enough temperature ranges to find out the Curie points of these samples.

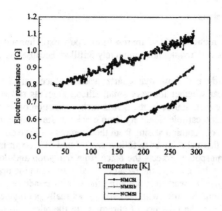

Figure 6. The variation of the electric resistance with temperature
of NMS1b, NMCS1 and NCMS2 showing the semimetallic behaviour of the samples. It is to remark the
noisy aspect of the curves corresponding to NCMS1 and NCMS2, which might be caused by the stress in
the crystals.

Figure 7 makes a summary on the magnetic behaviour under the mentioned conditions for NMS1b, NMS1t, NCMS1 and NCMS2. As probably expected, the saturation magnetisation presents the highest value for NMS1b, the sample with the true NiMnSb structure. The NMS1t is not too far from this, but definitely the composition of the sample influences on the magnetic properties as otherwise expected. On the contrary, the impure and not homogeneous sample NCMS1 grown in quartz has a weaker magnetisation. The surprise comes from the sample containing homogeneously distributed chromium-NCMS2, whose magnetic properties look closer to those of NMS1t.

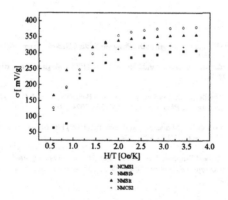

Figure 7. The magnetisation measurements at room temperature,
using the AC permeametre method

3.Discussions and conclusions

As we underlined in the introduction, these are results of a pure experimental work. We think that its main purpose has been touched, though not completely fulfilled, because there is a lot more to do in this direction.

The growth experiments do not show that quartz would induce oxidation of the NiMnSb batches. However, it is not yet clear whether the very small silicon amounts shown by the EDAX analyses in NCMS1 and NCMS2 have their source in the quartz-melt interaction or in the polishing process carried out at some stage with silicon carbide. The very deep cracks in the samples could still keep the particles after an ultrasound bath in our opinion arising from the attempts we have done with these samples.

In the same time, the best thermal regimes to follow are "2" and "3" shown in *Figure 1*. Here it is also to point out that nitrogen atmosphere protects the batch from oxidation and does not allow evaporation of the elements in the chamber, but changes the radial thermal gradient in the upper zone of the melt. That is why our very next purpose is to work to improving the growth conditions and keep going on with the regime "2", because the results obtained with NMS1b are not really negligible.

A separate subject is made by the presence of chromium in the alloy - in fact, we made the quaternary alloy NiCrMnSb, but it was incompletely worked out. The very first conclusion coming out from this experiment is that the main interaction happens between Cr and Ni, while the other components keeps their behaviour unchanged. This encourages us to go further with completely replacing nickel by chromium in a ternary alloy CrMnSb.

Both the quaternary and the forthcoming ternary alloys necessitate a profound and detailed study in any respect, because it might be interesting to learn what happens whith the Cr atoms in the NiMnSb lattice and whether this would help for instance to use one or both of them instead PtMnSb, which is a very expensive material. The up to now experiments show however that the magnetisation does not exibit a too low value in the case of NiCrMnSb.

Another point on which we have not really insisted is the treatment of the lattice dynamics in NiMnSb. We used the Raman measurements to check the antimony presence in the samples by a more surface sensitive means, that proved very efficient, but a complete study of the infrared and Raman behaviour of the NiMnSb alloy is necessary to complete the knowledge about this so actual though old material.

4. References

1. Castelliz, L. (1953) Eine ferromagnetische Phase im System Nickel-Mangan-Antimon, *Monatshefte für Chemie, Bd.* 82/6, 1059-1085

2. Kabani, R., Terada, M., Roshko, A. and Moodera, J. S. (1990), Magnetic properties of NiMnSb films, *J. Appl. Phys.* 67 (9), 4898-4900

3. de Groot, R. A., Mueller, F. .M., van Engen, P. G. and Buschow, K. H. J. (1983) New Class of Materials: Half-Metallic Ferromagnets, *Phys. Rev. Letters* 50(25), 2024-2026

4. Dinculescu, L., Logofatu, C., Mincu, N. and Iliescu, B., (1999) Fluoride crystals for IR optics grown by PVGF method, *J. Crystal Growth* 198/199, 999-1004

5. Philips, J., Chen, Y. and Dumesic, J. A. (1985) *Characterisation of Supported Iron-Oxide Particles Using Mössbauer Spectroscopy and Magnetic Susceptibility* ACS Symposium Series No 228, Marvin L. Diney and John L. Gland Eds.

Acknowledgments

The Royal Society and NATO are greatly acknowledged for partial support of this work during the postdoctoral fellowship of CEAG.
The authors are thankful to Professor O. Cerclier - Universite de Provence, Marseille, for invaluable help with XRD interpretation and calculations at some stages of this work.

ONE OF POSSIBLE APPLICATIONS OF HIGH-Tc SUPERCONDUCTORS

A.V.NIKULOV
*Institute of Microelectronics Technology and High Purity Materials,
Russian Academy of Sciences, 142432 Chernogolovka, Moscow
District, RUSSIA. E-mail: nikulov@ipmt-hpm.ac.ru*

Possible application of a result published recently is announced. According to this
result the chaotic energy of thermal fluctuation can be transformed in electric
energy of direct current in an inhomogeneous superconducting ring. Although the
power of this transformation is very weak this effect has optimistic perspective of
wide application because useful energy can be obtained from heat energy in the
equilibrium state. It is proved that this process is possible in spite of the prohibition
by the second law of thermodynamics. It is shown that the second law of
thermodynamics can be broken in a quantum system.

1. Introduction

Superconductors are used first of all as the conductors with zero resistivity. But
the superconductivity is not only the infinite conductivity. It is first of all a
macroscopic quantum phenomena. The long-range phase coherence can exist in a
superconducting state. The zero resistivity can be considered as one of the
consequences of the phase coherence. A non-zero velocity of the superconducting
electrons v_s can exist without a voltage in a state with long-range phase coherence
because

$$v_s = \frac{1}{m}\left(\frac{h}{2\pi}\nabla\varphi - \frac{2e}{c}A\right) \tag{1}$$

in this state. Here φ is the phase of the wave function; A is the vector potential; m is
the electron mass and e is the electron charge. Other consequences of the phase
coherence are the Meissner effect [1], the quantization of the magnetic flux in a
superconductor and the quantization of the velocity in a superconducting ring. In
the relation

$$\oint_l dl\, v_s = \frac{1}{m}\left(\frac{h}{2\pi}\oint_l dl\nabla\varphi - \frac{2e}{c}\oint_l dlA\right) = \frac{h}{m}\left(n - \frac{\Phi}{\Phi_0}\right) \tag{2}$$

obtained by the integration of (1) along a closed path l $n = \frac{1}{2\pi}\oint_l dl\nabla\varphi$ must

be an integer number since the wave function $\Psi = |\Psi|\exp(i\varphi)$ of the

183

R. Cloots et al. (eds.), Supermaterials, 183–191.

superconducting electrons must be a simple function. $\Phi_0 = hc/2e = 2.07 \; 10^{-7} \; G \; cm^2$ is the flux quantum; Φ is the magnetic flux contained within the closed path of integration l. If the $\varphi(r)$ function does not have a singularity inside l than n = 0. It is the cause of the Meissner effect. The $\varphi(r)$ function can have a singularity only if a nonsuperconducting hole is inside the superconductor. n can be a non-zero integer number in this case. If the superconductor thickness around the hole is big then a closed path l exists on which $v_s = 0$. According to (2) $\Phi = n\Phi_0$ in this case. In a opposite case of a superconducting ring with a narrow wall w the quantization of the velocity takes place. If the wall thickness is much smaller than the penetration depth of magnetic field w << λ then LI_s << Φ_0 and $\Phi \approx BS$. The permitted values of the velocity in a homogeneous ring are

$$v_s = \frac{h}{lm}\left(n - \frac{\Phi}{\Phi_0}\right) \approx \frac{h}{lm}\left(n - \frac{BS}{\Phi_0}\right) \qquad (3)$$

Here L is the inductance of the ring; $I_s = sj_s = s2en_sv_s$ is the superconducting current in the ring; n_s is the density of superconducting pairs; $l = 2\pi R$; R is the ring radius; B is the induction of the magnetic field induced by an external magnet; $S = \pi R^2$ is the ring area.

One of the consequences of the velocity quantization is the Little-Parks effect [2]. The $|v_s|$ tends towards a minimum possible value. Therefore according to (3) the energy of superconducting state of a ring (or a tube) and, as a consequence, the critical temperature of the ring

$$T_c(\Phi) = T_c\left[1 - \frac{\xi^2(0)}{R^2}\left(n - \frac{\Phi}{\Phi_0}\right)^2\right] \qquad (4)$$

depend in a periodic manner on the magnetic field value, with period Φ_0/S. The Little-Parks result [2] is the experimental evidence of the velocity quantization phenomena in superconductor. This phenomena is very important. The existence of forbidden states in superconducting ring means a possibility of a voltage induce by a change of the n_s value. Because the state with $v_s = 0$ is forbidden if the magnetic flux contained within the ring is not divisible by the flux quantum the superconducting current $I_s = sj_s = s2en_sv_s$ is not equal zero and changes with a n_s change. Consequently, according to the Faraday's law $lE = -d\Phi/dt$, a change of the superconducting pair density induces a voltage

$$E = -\frac{1}{l}\frac{d\Phi}{dt} = -\frac{1}{l}\left(Ls2e\frac{h}{lm}\left(n - \frac{\Phi}{\Phi_0}\right)\frac{dn_s}{dt} + L\frac{dI_n}{dt}\right) \qquad (5)$$

Here $\Phi = BS + LI_s + LI_n$ is the full magnetic flux inside the ring; I_n is the normal current in the ring; $I_n = lE/R_n$; $R_n = l\rho_n/s$ is the resistance of the ring in the normal state. After a sharp change of the n_s value the voltage decreases during the decay time L/R_n. The n_s value can change in a consequence of a change of the ring temperature. Thus, a small superconducting ring can be used as a thermo-electric generator. In the present report the most interesting case, an induction of direct voltage in an inhomogeneous ring, is considered.

2. Inhomogeneous Superconducting Ring as Direct-Current Generator

In a inhomogeneous superconducting ring with different density of superconducting pairs along the ring circumference the velocity has also different values. In the stationary state the normal current is equal zero and the superconducting current $I_s = sj_s = s2en_sv_s$ should be constant along the circumference. Consequently the velocity is higher in a section with lower n_s value. If a ring consists of two sections l_a and l_b ($l_a + l_b = l = 2\pi R$) with the density of superconducting pairs n_{sa} and n_{sb} the velocity (v_{sa} and v_{sb}) and the superconducting current in the permitted state according to (2) are equal

$$V_{sa} = \frac{h}{m} \frac{n_{sb}}{(l_a n_{sb} + l_b n_{sa})} (n - \frac{\Phi}{\Phi_0}) \qquad (6a)$$

$$V_{sb} = \frac{h}{m} \frac{n_{sa}}{(l_a n_{sb} + l_b n_{sa})} (n - \frac{\Phi}{\Phi_0}) \qquad (6b)$$

$$I_s = \frac{s2eh}{m} \frac{n_{sa} n_{sb}}{(l_a n_{sb} + l_b n_{sa})} (n - \frac{\Phi}{\Phi_0}) \qquad (6c)$$

Because the I_s value depends on the superconducting pair density of all ring section (see (6c)) it can be changed by a n_s change in any ring section. The potential difference dV/dl appears at an inhomogeneous change of the n_s value along the ring circumference. At an alternate change of the n_s value the average of the potential difference over a long time $<dV/dl>$ is equal zero if n_s keeps a non-zero value in whole ring. But the $<dV/dl>$ value can be not equal zero if a ring section is switched iteratively from the superconducting to normal state and backwards.

According to (6c) $I_s \neq 0$ if $n_{sa} > 0$, $n_{sb} > 0$ and $\Phi/\Phi_0 \neq n$. When (for example) the n_{sb} value changes from $n_{sb} = n_b > 0$ to $n_{sb} = 0$ (at $n_{sa} = const > 0$) the potential difference $V_b = R_{bn}I$ appears on the ring section l_b and the current I decreases exponentially from the initial value $I = I_s$ during the decay time L/R_{bn}. Because the stationary state with $I_s = 0$ is forbidden at $n_{sa} > 0$, $n_{sb} > 0$ and $\Phi/\Phi_0 \neq n$ the current in the ring increases to $I \approx I_s = (s2eh/m)(n_{sa}n_b/(l_an_b+l_bn_{sa}))(n - \Phi/\Phi_0)$ after the return of the l_b section to the superconducting state. A state with any integer number n is permitted. But the state with a minimum $|n - \Phi/\Phi_0|$ value has the greatest probability. Therefore the average of I_s is not equal zero at $\Phi/\Phi_0 \neq n$ and $\Phi/\Phi_0 \neq n + 0.5$. The average $<V_b>$ induced by the switching of the ring section from the superconducting to normal state and backwards is not equal zero also in this case.

If the energy difference between the lowest state with n and the next state with n+1 is bigger than k_BT and $t_n >> L/R_{bn}$, $t_s >> L/R_n$ then

$$<V_b>_t \approx \frac{s<n_{sb}>}{l_b n_{sa} + l_a <n_{sb}>} \frac{\Phi_0}{\lambda_{La}^2} (n - \frac{\Phi}{\Phi_0})Lf \qquad (7)$$

Here t_s and t_n are duration of the superconducting and normal states of the l_b section; $\lambda_L = (mc/4e^2n_{sa})^{0.5}$ is the London penetration depth of the l_a section; $f = 1/(t_s + t_n)$ is the frequency of the switching from the superconducting to the normal state. This result published in [3,4] means that a small superconducting ring can be used as direct-current generator [5] in which the heat energy can be transformed to the electric energy of the direct current.

If the temperature of the l_b section changes iteratively from $T_{min} = T_{cb} - \Delta T$ to $T_{max} > T_{cb}$ a part of the heat energy Q spent during a time t can be transformed in the electric energy $E_{d.c.} = tW_{d.c.} = t<V_b(t)I(t)>_t$. The efficiency $Ef = E_{d.c.} /Q$ of such generator is

$$Ef = \frac{R_b}{R_b + R_l} \frac{c_s}{c_s + c_n} \frac{\mu}{1+\mu}(1 - \frac{T_{min}}{T_{max}}) \tag{8}$$

Here R_l is the resistance of a load; $c_s = -T(dF_{GL}/dT)$ is the heat capacity of superconducting state; $c_s + c_n = c_b$ is the whole heat capacity; $\mu/(1+\mu)$ is the portion of the magnetic energy $E_L = LI_s^2/2$ in whole change on the superconductor energy. It has the maximum value $\mu_{max} = (32\pi^3/\kappa^2)(Ls/l_b^3)(n-\Phi/\Phi_0)^2$ at $T = T_c$. $\kappa = \lambda_L/\xi$ is the superconductor parameter introduced in the Ginzburg-Landau theory. The maximum efficiency, at $R_l \ll R_b$; $c_n \ll c_s$ and $\mu \gg 1$, is corresponded to the efficiency $Ef = (1 - T_{min}/T_{max})$ of the Carnot cycle considered first in 1824 year.

Thus, the superconducting ring does not differ qualitatively from any conventional heat engine if the switching of the ring section from the superconducting to normal state is caused by the temperature change. It's power is very small. Therefore any application of such generator would be senseless if the switching could be in the consequence only of the T change. But this switching can take place at an unaltered temperature in a consequence of the thermal fluctuation. If the ring section l_a and l_b have different values of the critical temperature $T_{ca} > T_{cb}$ then at an unaltered $T \approx T_{cb}$, $n_{sa} > 0$ always whereas the l_b section is switched by fluctuation from the superconducting to normal state and backwards. The direct voltage (7) should appear at both regular and chaotic switching. But the switching to the superconducting state should be simultaneous in whole section with lowest critical temperature. This simultaneous switching in the consequence of thermal fluctuation takes place in a mesoscopic section all sizes of which are not bigger than the superconducting coherence length ξ. At $T \approx T_{cb}$, the average resistance R_b of the l_b section $R_{bn} > R_b = R_{bn}t_n/(t_s+t_n) > 0$ because some times (t_n) $n_{sb} = 0$ (and therefore $R_b = R_{bn}$) and some times (t_s) $n_{sb} > 0$ (and therefore $R_b = R_{bn}$). Consequently, a direct voltage can appear in a mesoscopic inhomogeneous superconducting ring at an unaltered temperature corresponded to the resistive transition of the section with lowest T_c [3,4].

3. Possible Applications of Superconducting Ring System

The direct voltage appeared in the consequence of the thermal fluctuation can be used. Therefore the superconducting ring can be considered as electric power source in which the chaotic energy of thermal fluctuation is transformed to the direct current energy. The power $W = <V_b>^2 R_l/(R_l+R_b)^2$ can develop across a load with the resistance R_l. This power is obtained in a consequence of a regulating of the chaotic energy of the thermal fluctuation.

3.1. LIMITING POWER OF FLUCTUATION ENERGY REGULATING

Fundamental relation for limiting power of this regulating is determined by the characteristic energy of fluctuation k_BT and a time of a cycle which can not be shorter than $h/2\pi k_B T$ because of the uncertainty relation. Therefore $W < (k_B T)^2 2\pi/h$ in any case.

The power of one ring can not exceed

$$W_{max} \approx \frac{\mu}{1+\mu} \frac{16\pi k_B^2 T_c^2 Gi}{h} \qquad (9)$$

Here $Gi = (k_B T_c/H_c^2(0) l_b s)^{0.5}$ is the Ginzburg number; $H_c(0)$ is the thermodynamic critical field at $T = 0$. This value is very small: at $T_c = 100$ K, $(k_B T_c)^2 2\pi/h \approx 10^{-8}$ Wt and $8\pi Gi\mu/(1+\mu)$ value can not exceed 1 in any ring. Therefore the power even of high-Tc superconducting ring can not exceed 10^{-8} Wt. In order to obtain an acceptable power a system with big number of the rings should be used. The power $W = N^2 <V_b>^2 R_l/(R_l+NR_b)^2$ developed across the load R_l can be obtained by a system of N rings the sections of which are connected in series. This power has maximum value $W = N<V_b>^2/4R_b$ at $R_l = NR_b$. Because the power $<V_b>^2/R_b < 10^{-8}$ Wt at $T_c = 100$ K we should used a system with 4 10^8 HTSC rings in order to obtain the power 1 Wt.

Consequently in order to obtained the maximum possible power we should use HTSC rings with maximum possible $Gi\mu/(1+\mu)$ value.

3.2. REQUIREMENTS TO SUPERCONDUCTING RINGS AS DIRECT CURRENT GENERATOR

It is impossible now to give real optimum parameters of the ring system affording the maximum power of the fluctuation energy transformation. Additional theoretical and experimental investigations are necessary for this. I list only basic requirements which follow from the theoretical consideration.

Because the fluctuation induce the voltage in a narrow temperature region near T_{cb} and μ has maximum value at $T = T_{cb}$ all rings of the system should have the same critical temperature. A section of each ring should has a lower critical temperature T_{cb} than T_{ca} of other part of the ring. Sizes of this section should be enough small. They should not surpass considerably ξ/Gi_{3D}. Here Gi_{3D} is the Ginzburg number of three-dimensional superconductor. The ξ/Gi_{3D} value of known HTSC is very small. Therefore methods of nano-technology are necessary for the making of the ring system. The l_a section with higher T_c can be longer considerably than l_b. The μ_{max} value increases with l_a because the ring inductance L is proportional to l. But the l_a should not be too long. At big l_a/l_b value the transition to the superconducting state of the l_b section becomes first order if $| n - \Phi/\Phi_0|$ is not small [3]. The fluctuation can not switch the l_b section from the superconducting to normal state and backwards if $l_a s$ is too big. I suppose the optimum diameter of HTSC ring should be approximately 1 μm. The system of 10^8 rings with diameter 1 μm can go in an area 1 cm².

3.3. POWER SOURCE AND MICRO-REFRIGERATOR WITHOUT EXPENSE OF AN EXTERNAL ENERGY

Although the power of the proposed direct-current generator is very weak and it's making is very difficult it has optimistic perspective of application. The proposed power source does not need an expense of an external energy. Moreover it can be used simultaneously as a micro-refrigerator [6]. The chaotic heat energy is transformed in the ordered electric energy in the proposed system of superconducting rings. The ordered energy turns back into the heat energy during an utilization. Therefore the proposed power source can work anyhow long time without an expence of any fuel. Such power source may be especially useful in self-cotained systems.

4. Why the chaotic heat energy can be ordered in a quantum system

The perspective of considered application is connected with the possibility of the violation of the second law of thermodynamics which should take place [7] if the result published in [3,4] is correct. In a consequence of this circumstance most scientists can not believe that the result [3,4] can be correct. Below I try to prove that the proposed application of the HTSC rings can be real because the second law of thermodynamics can be broken in a quantum system. I explain also why the voltage with direct component can exist in the superconducting ring section.

4.1. A QUANTUM FORCE

According to (7) the direct voltage $<E_b> = <V_b>/l_b$ can be induced in the ring section by it's iterative switching from the superconducting to normal state and backwards. Because

$$E = -\nabla V - \frac{dA}{dt} \qquad (10)$$

and consequently $\oint_l < E > dl = < V_a > + < V_b > = - < \frac{d\Phi}{dt} > = 0$ the

direct voltage $<E_a> = <V_a>/l_a = -<V_b>/l_a$ exists in the superconducting section l_a if it exists in the switched section l_b. The superconducting pair in the l_b section should accelerate continually (because $mdv_s/dt = -eE$) and consequently the result [3,4] can not be correct if only electric force $F_e = -eE$ acts on electrons. But if we assumed this we would conclude that the Meissner effect is impossible. If $2mdv_s/dt = -2eE$ then $2mdv_s/dt + 2eE = d(2mv_s - 2eA/c)/dt = 0$ and consequently the $mv_s- e\Phi/lc$ value should be constant in time. The magnetic flux should not change at the transition to the superconducting state. But the superconductivity is a macroscopic quantum phenomena. Therefore the superconducting state differs from the state with infinite conductivity. The superconducting pairs can accelerate against the electric force at the transition to the superconducting state as it takes place at the Meissner effect. Consequently an additional force can act on the electrons at the transition to the superconducting state. Because this force is caused by the quantization of the electron velocity I will call it as quantum force F_q.

The quantum force acts at the closing of the superconducting state in the ring. At the transition of the l_b section from normal $n_{sb} = 0$ to superconducting state with $n_{sb} = n_{sa}$ the velocity changes from $v_s = 0$ to $v_s = (e/mcl)(n\Phi_0 - \Phi)$ and the magnetic flux inside the ring Φ changes from $\Phi = BS$ to $\Phi = BS + LI_s$. The mv_s-$e\Phi/cl$ value changes by $(e/cl)(n\Phi_0 - \Phi) + (e/c)LI_s = (e/cl)(1 + sL/\lambda_L^2 l)(n\Phi_0 - \Phi)$.

Consequently at this transition $\int dtF_q = \int dt(m\frac{dv_s}{dt} - \frac{e}{cl}\frac{d\Phi}{dt}) = (e/cl)(1 + sL/\lambda_L^2 l)(n\Phi_0 - \Phi)$.

The quantum force F_q acts only at the transition from the normal to the superconducting state. At the transition of the l_b section to the normal state the v_s value in the l_a section decreases in a consequence of the electric force $F_e = -eE$. The electric field E is induced by the deceleration of electrons in the l_b section in a consequence of the energy dissipation. Thus, the superconducting pairs in the l_a section are accelerate by the quantum force F_q and are retarded by electric force F_e at the switching of the l_b section. They are not accelerated continually because $<F_q + F_e> = 0$. The $<F_e>$ is not equal zero at the switching of the l_b section because $<F_q> \neq 0$.

4.2. VIOLATION OF THE SECOND LAW OF THERMODYNAMIC

According to the Carnot's principle (1824 year), which may be considered as a first formula of the second law of thermodynamic, in the equilibrium state the heat energy can not be transformed to a work. The transformation of the fluctuation energy to electric energy of the direct current at an unaltered temperature considered above contradicts to this principle. It contradicts also to the Clausius's formula of the second law of thermodynamic (1850 year) if a temperature of the load is higher than the ring temperature. It contradicts to the Thomson's formula (1850 year) if the load is an electricmotor. Nevertheless I state that the results [3,4] is correct because the second law of thermodynamic can be broken in a quantum system. I understand it is very difficult convince anyone that this claim can be correct. Most persons are fully confident that the second law of thermodynamic can not be broken. Although few authors fight against this law (see [8]).Below I try to explaine because my claim is correct.

The violation of the second law of thermodynamic means that the heat energy can be transformed to the work (in a regular energy) in the equilibrium thermodynamic state. That is the work is posssible without an expence of any fuel. Because the regular energy is transformed to the chaotic heat energy at the work, the work can be put out anyhow long time. The work is $A = \int FdX$. Here F is an abstract force; X is an abstract coordinate. Because $dX = vdt$, $A = \int Fvdt$.

Consequently the violation of the second law of thermodynamic means that the average value $<Fv> = \int_0^T Fvdt / T$ by the infinite time T can be not equal zero. Any closed thermodynamic system has come to the equilibrium state with maximum entropy during any time. Therefore the average $<Fv>$ should be considered in the equilibrium state. In the last centure the second law of thermodynamic was interpreted as the impossipility of any dynamic process in the equilibrium state. In this interpretation $<Fv> = 0$ because $F = 0$.

In the beginning of our century physicists have understood that the dynamic processes take place in the equilibrium state in a consequence of the fluctuation. Therefore the fluctuation was considered as the breakdown of the second law of thermodynamic in that time [9]. But the fluctuation dynamic process is chaotic. Therefore $<Fv> = <F><v>$ and the work is equal zero if $<F> = 0$ or $<v> = 0$. In any conventional heat engine the force F and the velocity v are correlated. This correlation is achieved by an unequilibrium state. The entropy $S = Q/T$ increases at the work of a conventional heat engine because the heat energy Q is transferred from a hot body with T_1 to a cold body with T_2 during this process and $\Delta S = Q/T_2 - Q/T_1 > 0$ if $T_1 > T_2$. Numerous attempts to devise a correlated process with $<Fv> \neq <F><v>$ in the equilibrium state were unsuccessful for the present [8].

But the second law of thermodynamic can be broken at $<Fv> = <F><v>$ if $<F> \neq 0$ and $<v> \neq 0$. It is obvious that the motion should be circular in this case in order the work can be put out anyhow long time. In any classical (no quantum) system $<v> = 0$ because all states are permitted. The probability of a state P is proportional to exp-$(E/k_B T)$. The energy E of a state is function of v^2 in a consequence of the space symmetry. Therefore the probability of the state with v is equal the one with -v. Consequently $<v> = 0$ if all states are permitted. In a quantum state no all states are permitted. Therefore $<v>$ can be not equal zero in some quantum system. The superconducting ring with $\Phi/\Phi_0 \neq n$ and $\Phi/\Phi_0 \neq n + 0.5$ is one of such system. At $\Phi/\Phi_0 \neq n$ and $\Phi/\Phi_0 \neq n + 0.5$ the permitted states with the opposite directed velocity have different values of the kinetic energy $E_{kin} = mv_s^2/2$. For example at $\Phi/\Phi_0 = 1/4$ the lowest permitted velocities in a homogeneous superconducting ring are equal $v_s = -h/ml4$ at n=0 and $v_s = 3h/ml4$ at n=1. The kinetic energy of these states differ in 9 times. Therefore the thermodynamic average of the velocity is not equal zero.

The average force $<F>$ can be not equal zero at the switching of the ring section from the superconducting to normal state and backwards because the quantum force F_q acts at the closing of the superconducting state if $\Phi/\Phi_0 \neq n$. This force acts in the same direction (more exactly, has higher probability in one direction) if $\Phi/\Phi_0 \neq n + 0.5$. This direction coincides with the velocity direction. Therefore $<F_q><v> > 0$. This means that a part of the fluctuation energy can be ordered at the closing of the superconducting state in the ring. In order to switch the ring section to superconducting state the fluctuations should spend a energy $E_{kin} = sln_s mv_s^2/2$ for the acceleration of the superconducting electrons and a energy $E_L = LI_s^2/2$ for the induce of the magnetic flux $\Delta\Phi = LI_s$ of the superconducting current. This energy is no chaotic because the velocity has higher probability in one direction. The regulated energy is dissipated in the l_b section at it's transition to the normal state. But we can use a part of this energy for a work if we put load a R_l on the voltage appeared on the l_b section as it is proposed above.

Thus, the new application of HTSC proposed in the present work is real although it contradicts to the second law of thermodynamic.

5. Acknowledgments

I am grateful Jorge Berger for the preprint of his paper and for stimulant discussion. This work is supported by the International Association for the

Promotion of Co-operation with Scientists from the New Independent States (Project INTAS-96-0452) and National Scientific Council on Superconductivity. I thank these organizations for financial support.

6. References

1. Shoenberg, D. (1952) *Superconductivity*, Cambridge.
2. Tinkham, M. (1975) *Introduction to Superconductivity*, McGraw-Hill Book Company, New-York.
3. Nikulov, A.V. and Zhilyaev, I.N. (1998) The Little-Parks effect in an inhomogeneous superconducting ring, *J. Low Temp. Phys.* **112**, 227-236; http://xxx.lanl.gov/abs/cond-mat/9811148.
4. Nikulov, A.V. (1999) Transformation of Thermal Energy in Electric Energy in an Inhomogeneous Superconducting Ring, in *Symmetry and Pairing in Superconductors,* Eds. M.Ausloos and S.Kruchinin, Kluwer Academic Publishers, Dordrecht, 373-382; http://xxx.lanl.gov/abs/cond-mat/9901103.
5. Nikulov, A.V. (1999) A superconducting mesoscopic ring as direct-current generator, *Abstracts of NATO ASI "Quantum Mesoscopic Phenomena and Mesoscopic Devices in Microelectronics"* Ankara, Turkey, p.105-106
6. Nikulov, A.V. (1999) A system of mesoscopic superconducting rings as a microrefrigerator, *Proceedings of the Symposium on Micro- and Nanocryogenics*, Jyvaskyla, Finland, p.68.
7. Nikulov, A.V. (1999) Violation of the second thermodynamic law in a superconducting ring, *Abstracts of XXII International Conference on Low Temperature Physics,* Helsinki, Finland, p.498
8. Berger J. (1994) The fight against the second law of thermodynamics, *Physics Essays* **7**, 281-296.
9. Smoluchowski, M. (1914) Gultigkeitsgrenzen des zweiten Hauptsatzes der Warmetheorie, in *Vortrage uber kinetische Theorie der Materie und der Elektrizitat (Mathematische Vorlesungen an der Universitat Gottingen, VI).* Leipzig und Berlin, B.G.Teubner, p.87-105

ORDER PARAMETER SYMMETRY
AND LOW-TEMPERATURE ASYMPTOTICS
FOR MESOSCOPICALLY NONHOMOGENEOUS
SUPERCONDUCTORS

Low-temperature asymptotics for superconductors

ALEXANDER M. GABOVICH AND ALEXANDER I. VOITENKO
Crystal Physics Department, Institute of Physics, NASU
prospekt Nauki 46, 03022 Kiev-22, Ukraine

1. Introduction

The electron-phonon nature of superconductivity in metals and alloys had been perceived even before the derivation of the Bardeen-Cooper-Schrieffer (BCS) theory based on the Cooper pairing concept [1]. The superconducting order parameters (OP's) for all objects discovered by that time turned out to be spin-singlet and isotropic (*s*-wave). All early attempts to introduce either non-phonon mechanisms or order parameters of alternative symmetries [2, 3, 4, 5, 6] gained almost all recognition over a long period.

The situation changed in 80-*s* when a number of superconductors, widely called "exotic", were found [7, 8, 9]. Among these one can mention organic low-dimensional metals, magnetic superconductors, in particular, heavy-fermion compounds, high-T_c (at that time) solid solutions $BaPb_{1-x}Bi_xO_3$ (BPB), inorganic polymer $(SN)_x$, tungsten bronzes, $SrTiO_3$, etc. Their thermodynamic, electromagnetic, and transport properties did not follow the simple corresponding state laws appropriate to the BCS scheme [1]. Now the deviations from these laws do not seem surprising, the more so after the discovery of the ramified oxide families with really high T_c's [10, 11, 12]. Presumably, it is precisely high T_c's that make features of various specific substances detectable [13, 14].

Nevertheless, the explanations based on the traditional *s*-wave description of superconductors go out of fashion, especially for cuprates, where *d*-wave spin-singlet pairing is almost unanimously considered as the only possible or at least as the dominant one [15, 16, 17, 18, 19]. Moreover, the "day of unconventional superconductors" with OP's of *d*-, *p*-, *f*-, and mixed symmetries has been proclaimed also for all heavy-fermion com-

193

R. Cloots et al. (eds.), Supermaterials, 193–212.

pounds and organic superconductors [20], Sr_2RuO_4 [21], and even Laves phase $CeCo_2$ [22, 23]. Really, good grounds exist for such conclusions (see also reviews [18, 19, 24]). The nontraditional explanations of the experimental data are, however, not at all unavoidable since there is a lot of evidence matching more closely the s-wave picture. The apparent ambiguity of the existing interpretations is due to the unfortunate absence of "the" theory of superconductivity making allowance for the nontrivial background normal-state properties [25, 26, 27, 28]. Thus, phenomenological analysis and critical insight remain so far the most important tool of investigation.

Below we express our point of view on the subject concerned being confined mainly to copper oxides. After a brief review of the controversies over the OP symmetry an original reconsideration of the low-temperature (LT) superconducting properties is presented.

The famous and already rewarded phase-sensitive experiments are usually judged to be the most trustworthy illustration of the d-wave pairing in hole-doped cuprates [15, 16, 29, 30, 31]. Nevertheless, a suppression of the ordinary s-wave OP at twin boundaries, the flux trapping there or in the corners, and meanderings of the grain boundaries can reproduce [32, 33, 34] the observed half-flux spontaneous magnetization or π-shifts of the phases across c-axis junctions [29, 30]. Also the recent phase-sensitive experiments, namely, measurements of the nonlinear Meissner effect [35, 36] on untwinned $YBa_2Cu_3O_{7-x}$ (YBCO) and the twist angle independence of the c-axis critical current density observed in c-axis twist bicrystals of $Bi_2Sr_2CaCu_2O_{8+x}$ (BSCCO) [37] better agree with the isotropic pairing than with an anisotropic one.

On the other hand, the emergence of the dielectric gap on the nesting Fermi surface (FS) sections due to the charge-density wave (CDW) formation may mimic the superconducting pseudogap above the critical temperature T_c and severely hamper various measurements of the superconducting gap below T_c [32, 33, 38, 39]. For example, the predicted current-voltage characteristics (CVC's) asymmetricity for junctions involving CDW superconductors with s-type pairing [38] resembles that of the ab-plane tunnel CVC's for BSCCO [40]. The same can be said about the persistence of the smeared pseudogap features in this experiment and the measurements [41], so that the CDW's rather than the d-wave scenario with V-shaped conductance show themselves. There are also other data obtained by tunnel and point-contact spectroscopies which are inconsistent with d- wave picture. In particular, the investigations [42, 43] of inelastic Cooper pair tunneling for various phases of the Bi-Sr-Ca-Cu-O system clearly demonstrated the existence of the Riedel-like singularity and the subsequent steep reduction of the Josephson current inherent to BCS isotropic superconductors [44], whereas the d-wave picture lacks such a threshold behavior [45]. The indications of the d-wave inconsistency with measured photoexcited relaxation dynamics in YBCO were also found [46]. Moreover, such time-domain spectroscopy reveals the profound distinctions between the temperature, T, independent pseudogap and T-dependent superconducting gap simultaneously existing in the sample [47]. Pseudogaps, observed by other methods [48], including Raman, infrared, and angle-resolved photoemission measurements, according to our point of view [38, 49], should be also identified with the CDW gaps.

The direct evidence was obtained for the s-wave pairing in YBCO and $YbBa_2Cu_3O_{7-x}$ using tunnel and point-contact measurements [50] as well as in BSCCO using tunnel spectroscopy [41, 51], where clear classical gap features were seen. It is necessary also to indicate the wave vector independence of the superconducting pairing gap features in $La_{2-x}Sr_xCuO_4$ (LSCO), measured by the inelastic neutron scattering, so that the experimenters were not able to reconcile their results with the mainstream of the d-wave interpretation [52].

There are data testifying that for a number of oxides the mixing of the OP occurs with different symmetries. First, the c-axis Josephson tunneling between definitely s-wave Pb electrode and purported d-wave YBCO [30, 53, 54] or BSCCO [55] one manifests a conventional s-wave Ambegaokar-Baratoff behavior but with the lowered critical current amplitude. Second, microwave conductivity measurements in YBCO appeared inconsistent with either single s- or single d-like OP fitting [56]. Third, electronic Raman scattering experiments reveal the OP anisotropy, gap zeros, but also deviations from the pure d-wave picture. Specifically, in LSCO the inherent to d-symmetry ω^3 shape of the B_{1g} peak is absent for all doping levels [57]. In disordered $Bi_2Sr_2Ca(Cu_{1-x}Fe_x)_2O_{8+y}$ the OP nodes are shifted from the positions imposed by pure d-wave symmetry [58, 59]. It was predicted that the amount of the s-wave component can be determined, in principle, from experiments of this kind [60]. The mixing between OP's with different symmetries in high-T_c oxides is allowed by the orthorhombic crystal structure in the majority of these compounds [15].

At last, the possibility of the d-symmetry pairing is supported by a lot of experiments demonstrating power-law T- dependences of various thermodynamic and transport properties far below T_c [15, 16, 32, 33, 61, 62] , instead of the almost exponential gap-driven BCS ones [1]. These results are usually viewed as an evidence of gap point or line nodes on the FS. In this paper we analyze the current situation with relevant experimental data and their theoretical description, bearing in mind the observed noticeable amounts of s-wave contributions to the oxide OP's. It is shown that power-law characteristics are compatible even with the pure s-like superconductivity if one takes into account the wide distribution of the superconducting OP magnitudes always existing in complex nonhomogeneous structures of high-T_c oxides [63, 64, 65, 66], in contrast to simple metals and alloys. That is why OP's of various symmetry, including mixed ones, remain possible candidates for the true Cooper pair orbital wave function in oxides. In this connection, first, one can mention the controversy over the so-called "paramagnetic Meissner effect" observed both in cuprates and Nb [32, 33, 67, 68]. The generic nature of this phenomenon was recently demonstrated for Al disks and attributed to the surface superconductivity [69]. Second, a power-law electronic contribution to the heat capacity of Nb$_3$Sn observed by different groups [70, 71, 72] instead of the exponential one remained a puzzle since (but can be easily explained in the framework of our approach). Thus, the necessity of further Cooper pair symmetry investigations persist to be on agenda with an emphasis on comparison between analogous properties of high- and low-T_c oxides and other superconductors proven to be conventional s-wave objects.

From the results presented in Sec. 3 it becomes clear that the allowance for the OP distribution results in the observed T-dependences of various properties different from those inherent to the homogeneous superconductors whatever the type of pairing. Since the change is more remarkable for the s-wave pairing (from the exponential dependence to the power-law one), in this paper we confine ourselves to the isotropic case. The generalization to the more complicated forms of the OP is straightforward and will be considered elsewhere.

2. BCS asymptotics and experimental situation

From the original weak-coupling BCS theory it comes about that whatever the behavior of the certain thermodynamic or transport electron characteristics in the T_c neighborhood, their low-T asymptotics are always proportional to the exponential factor $\exp\left(-\frac{\Delta_0}{T}\right)$, where Δ_0 is the value of the isotropic s-gap at $T = 0$ [1]. (Hereafter we assume $k_B = \hbar = 1$.) In particular, the specific heat asymptotics has the form

$$C_s(T) = N(0) \left(\frac{2\pi\Delta_0^5}{T^3}\right)^{1/2} \exp\left(-\frac{\Delta_0}{T}\right), \qquad (1)$$

where $N(0)$ is the electron density of states at the Fermi level. The temperature variation of the London penetration depth $\lambda_L(T)$ is governed by the superconducting electron density and has the form

$$\delta\lambda_L(T) = \frac{\lambda_L(T) - \lambda_L(0)}{\lambda_L(0)} = \left(\frac{\pi\Delta_0}{2T}\right)^{1/2} \exp\left(-\frac{\Delta_0}{T}\right). \qquad (2)$$

Some important transport properties of BCS superconductors can also be described analytically for $T \ll T_c$, e.g., electron thermal conductivity

$$\kappa_e(T) = \frac{2N_e\tau_{\text{tr}}}{m_e} \frac{\Delta_0^2}{T} \exp\left(-\frac{\Delta_0}{T}\right) \tag{3}$$

and the ratio of the ultrasound attenuation coefficients in superconducting, α_s, and normal, α_n, states

$$\delta\alpha(T) = \frac{\alpha_s}{\alpha_n} = \frac{2}{1 + \exp\left(\frac{\Delta_0}{T}\right)} \approx 2\exp\left(-\frac{\Delta_0}{T}\right). \tag{4}$$

Here N_e is the normal electron density, τ_{tr} is the transport collision time, m_e is the electron mass. The exponential T-dependences for $T \ll T_c$ in the BCS theory are also inherent to the nuclear spin-lattice relaxation rate T_1^{-1} and the electromagnetic absorption coefficient [1].

The observed characteristics of high-T_c oxides reveal quite different behavior, which is usually considered as a basis for the adoption of the d-wave concept [8, 15, 16]. In particular, for YBCO the specific heat $C_s \propto T$ [32, 33], although the recent experiment [62, 73] shows additional contribution $\propto T^2$ and Schottky anomalies $\propto T^{-2}$, making the whole picture uncertain. At the same time, for the d-wave gap function the results would have been proportional to T^2 for hexagonal or T^3 for cubic lattices [74].

For $\delta\lambda_L(T)$ the experimental data are quite ambiguous. For nominally pure YBCO samples linear dependences on T are observed [75], whereas for Zn- and Ni-doped as well as nonhomogeneous crystals $\delta\lambda_L \propto T^2$ (Refs. [75, 76, 77, 78, 79]) in a formal accordance with the theory of d-wave superconductors, dirty [16] or with surface-induced Andreev bound states [79]. On the other hand, there are data showing two-gap low-T asymptotics of $\delta\lambda_L(T)$ in YBCO [80]. The authors of Ref. [80] claim that this dependence is intrinsic, whereas the results of Ref. [75] are due to a non-uniform sample oxygenation. In mercury-based cuprates $HgBa_2Ca_{n-1}Cu_nO_{2n+2+x}$ recent measurements reveal that the in-plane penetration depth $\lambda_{ab} \propto T$, whereas the out-of-plane $\lambda_c \propto T^5$ for $n = 1$ and $\propto T^2$ for $n = 3$ [81]. Numerous measurements were made for different disordered cuprates demonstrating the general trend of $\lambda_{ab}(T)$ weakening and $\lambda_c(T)$ strengthening for stronger disorder [61, 82]. The most unusual dependence was observed in $YBa_2Cu_4O_8$, namely, $\lambda_{ab}(T)$ and $\lambda_c(T)$ are proportional to $\sqrt{T}$ up to $0.4T_c$ [83]. At the same time, for electron-doped oxide $Nd_{1.85}Ce_{0.15}CuO_{4-\delta}$, which always reveals only s-wave features, $\delta\lambda_L(T)$ follows the law (2) [15]. On the other hand, the electron-doped $Pr_{2-x}Ce_xCuO_4$ and the hole-doped $La_{2-x}Sr_xCuO_4$ demonstrate the similarity of the resistive properties [84]. Therefore, the striking difference between the responses to the magnetic field of the electron- and hole-doped cuprates seems quite surprising.

Unfortunately, it is hard to extract the electronic thermal conductivity component κ_e from the experiment due to the complex mutual action of electrons, phonons, and impurities [17, 85, 86]. Nevertheless, the existing experiments indicate that $\kappa_e \propto T$ in Zn-doped YBCO [87] and below $T_c^* = 200\,\mathrm{mK}$ in $Bi_2Sr_2Ca(Cu_{1-x}Ni_x)_2O_8$ [88]. The ultrasound attenuation coefficient α_s also exhibits a power-law T^M decrease for $T \ll T_c$ both for YBCO [89, 90] and $La_{1.8}Sr_{0.2}CuO_{4-x}$ [89] with a large scatter of the exponent M for each substance. As for the nuclear spin-lattice relaxation rate T_1^{-1}, it demonstrates power-law dependences with $1 \le M \le 3$ [16].

All this experimental material shows that the universal dependence for any phenomenon discussed does not exist. Moreover, the agreement with the theories based on the underlying hypothesis of the gap function is superficial. Even with additional assumptions being made, the equality between experimental and theoretical power-law exponents still can not be ensured. What is more important, the power-law asymptotics can be hardly reconciled with the undoubtedly observed s-wave components of the OP. At the same time, while analyzing local structures observed in the nonstoichiometric superconducting and nonsuperconducting oxides [63, 64, 65, 66, 91, 92], we made a conclusion that there should be another solution to the problem. Our approach starts from the assumption of a wide distribution of Δ values in the bulk of the samples at each $T < T_c$.

The inhomogeneities leading to the spread of Δ magnitudes over the sample may be of different nature. As the possible driven forces of these structural and/or electronic domains in high-T_c oxides one should mention (i) composition irregularities, especially the inherent disorder in oxygen vacancy positions, observed, e.g., for BPB [9, 93], $La_{2-x}[Sr(Ba)]_xCuO_{4-y}$ [93, 94], and YBCO [65, 66, 93, 95, 96]; and (ii) the phase separation of the electronic origin with impurity atoms frozen because of the kinetic barriers [97, 98]. In oxides both mechanisms apparently act together [91, 96, 99].

Vacancy disorder comprise point-like defects. An attempt to allow for such irregularities while studying nonhomogeneous superconductors was made in Ref. [100]. The relationship between this theory and ours will be discussed in the next Section.

The inhomogeneities attributed above to the second group are of typical sizes exceeding the coherence length, the latter being extremely small in cuprates. The experimental evidence exists of the minority phase domains in LSCO being as large as several hundred Angströms in size [99]. For YBCO the X-ray and neutron diffraction measurements supplemented by the lattice gas Monte-Carlo simulations revealed not only tetragonal and ortho-I phases with the long range order but also a rich variety of structural phases with anisotropic correlation lengths of mesoscopic size [96]. The domain finiteness preserved even after annealing, and kinetic barriers turned out to be large enough to secure the logarithmic time ordering. The crystal field neutron spectra of $ErBa_2Cu_3O_{7-x}$ [101, 102] and the Raman spectra of YBCO [103], which reflect the local region properties, also revealed oxygen structure domains, indicating the phase separation and the percolation character of conductivity and superconductivity. In Ref. [104] it was pointed out as well that the percolative network of intermediate size hole-induced polarons (clusters) may lead to the difference between local and global (crystal) symmetry. Therefore, the pseudogap properties of the underdoped cuprates may originate not from the long-range charged stripe structure but from a randomly distributed local patches [105].

In contrast to YBCO, the electron-doped oxide $Nd_{2-x}Ce_xO_{4-y}$ (NCCO) is a random alloy [106]. Such an atomic-scale disorder may prevent the formation of structural domains, thus making our hypothesis of averaging inapplicable in this case. On the other hand, the in-plane coherence length in NCCO is $\xi_{ab} \approx 70 - 80\,\text{Å}$, which exceeds substantially $\xi_{ab} \approx 10 - 15\,\text{Å}$ in YBCO [106].

It is essential to take into account that the situation discussed is even more intricate, since the samples made of these substances and other high-T_c oxides are not necessarily at equilibrium [63, 93]. A wide time-varied distribution of Δ over the sample may explain the well-known puzzle in BPB [9, 107]. Namely, Meissner effect was observed, although adiabatical calorimetry did not reveal any traces of an anomaly at T_c [107]. Later it was shown that freshly sintered ceramics exhibits clear specific heat anomaly near T_c and the exponential drop at $T \ll T_c$ [9]. However, the storage during $1\frac{1}{2}$ months leaving the diamagnetic Meissner response almost unaltered leads (i) to wiping the anomaly out and (ii) to nonexponential $C_s(T)$ dependence below T_c. It is the diffusion process in the oxygen octahedra that seems to be responsible for both effects. Aging effects with the unattainable equilibrium state and the characteristic relaxation period to a scale of days manifest themselves in the specific heat measurements for $SmBa_2Cu_3O_{7-x}$ [108]. The nuclear magnetic resonance experiments in $RBa_2Cu_3O_{6+x}$ (R = Tm, Y) also show a drastic temporal decrease of the samples' superconductivity fraction with no change in T_c [109]. Relaxational processes directly affecting T_c's were observed for YBCO [110, 111], $Tl_2Ba_2CuO_{6+x}$ [112], and $HgBa_2CuO_{4+x}$ [113]. Another important source of the Δ scatter is the CDW emergence in superconducting oxides [9, 32, 33, 38, 49, 65, 66] . All factors listed above, taken together or separately, may be responsible for the transition from the primordial exponential to the power-law behavior of the quantities under consideration thus hiding the s-wave contribution to the low-T properties.

It should be noted that structural mesoscopic nonhomogeneities induced by the local strains in the martensitically transformed samples also reveal themselves in A15 compounds [114], with one of its representatives, Nb_3Sn, possessing very small $\xi \approx 28$ Å (see, e.g., Ref. [115]) and showing for the heat capacity the unconventional T-linear term in the superconducting state [70, 71, 72, 114]. For this substance the large scatter in the energy gap, so that $2\Delta/T_c = 0.2 - 4.8$, was observed directly by the tunneling measurements [115, 116, 117, 118, 119, 120] .

3. Theory

The key idea of the theory is that not only a polycrystalline but even a single crystal superconducting oxide sample can be considered as *mesoscopically nonhomogeneous,* i. e., consisting of domains. The approach can be applied both to pure *s*- or *d*- as well as to mixed-symmetry *(s+d)*-superconductors. We, however, restrict ourselves in this article only to *s*-like OP. The reason is that only the nodeless isotropic case will be affected qualitatively by the spatial averaging described below. Thus, either for A15 compounds where the superconductivity is of the *s*-type or for cuprates where it has predominantly *d*-symmetry, the exponential contributions to thermodynamic or transport properties may originate only from the *s*-OP component. Its observed disappearance is of the main significance. On the other hand, an OP with line or point nodes leads to the power-law low-T asymptotics even before the averaging procedure. Of course, the spatial averaging may change the power exponent but this crossover is more difficult for the unambiguous interpretation.

The assumed domain structure is supposed to be T-independent, with each domain having the following properties:

(A) at $T = 0$ it is described by a certain superconducting OP Δ_0;

(B) up to the relevant critical temperature $T_{c0}(\Delta_0) = \frac{\gamma}{\pi}\Delta_0$, where $\gamma = 1.7810\ldots$ is the Euler constant, it behaves as a true BCS superconductor, i.e., the temperature dependence $\Delta(T)$ of the superconducting OP is the Mühlschegel function $\Delta(T) = \Delta_{BCS}(\Delta_0, T)$; any property P under investigation is characterized in this interval by the function $P_s(\Delta, T)$;

(C) at $T > T_{c0}$ it changes into the normal state, and the relevant property is $P_n(T)$.

At the same time, the values of Δ_0 scatter for various domains. The current carriers move freely across domains and inside each domain acquire the respective properties. Thus, possible proximity effects resulting in the correlation of the properties of adjacent domains are neglected. The current carrier density is assumed constant all over the sample, so transient processes are excluded from consideration.

The averaging procedure considered below requires (i) the effective sample size L to be much larger than the mean size of the domains d_{mean} and (ii) the size of each domain d_i to be larger than the relevant coherence length ξ_i. The first condition is needed to regard the superconductor macroscopically homogeneous. The second one stems from the property (B) indicated above. In the opposite case, when $d_i \ll \xi_i$, we are led to the lattice model of superconductor with a local atomic disorder [121, 122]. Such a model was applied to the description of YBCO in Ref. [100]. Strictly speaking, the authors tried to overcome the conclusion of the Anderson theorem [1] that static impurities weakly influence superconducting properties of isotropic metals. They showed that only for an anomalously great dispersion W of the site order parameters Δ_i it is possible to obtain the gapless-like behavior of the quasiparticle density of states. Considering the condition $W \gg \Delta_i^{\max}$ for the maximal quantity $\Delta_i^{\max}$ very improbable, the cited authors, in order to explain the experimental data, argued that YBCO is a pure d-wave object. However, this statement seems inconclusive because the model with different local pairing amplitudes on the neighboring sites seems unrealistic even for high-T_c oxides with small coherence lengths, so that one can hardly infer OP symmetry on its basis. In essence, the domain size there is comparable to that of the elementary cell. But in this limiting case we go beyond the scope of the BCS picture based on the long-range character of the phonon-induced interaction between electrons [1]. Even if such a model is adopted for cuprates, it leads to the Bose-condensation picture [123] rather than to the Cooper pairing. On the other hand, if we accept the Hubbard or $t - J$ models at the very beginning the choice in favor of the specific pairing mechanism and the anisotropic, e.g., d-OP becomes almost inevitable making any further judgement unnecessary. In contrast, the actually admitted condition $d_i > \xi_i$ is fully in line with our basic concept of averaging.

Under the assumed conditions, the current carrier liquid involves normal, $\rho_n(T)$, and superconducting, $\rho_s(T)$, fractions with $\rho_n(T) + \rho_s(T) = 1$, the superconducting fraction being multicomponent. Each superconducting component corresponds to domains with a certain Δ_0. They possess the properties (A), (B), (C) mentioned above. The superconducting fraction at $T = 0$ can be described by a distribution function $f_0(\Delta_0)$ in the interval $0 \leq \Delta_0 \leq \Delta_0^{\max}$:

$$\rho_s(0) = \int_0^{\Delta_0^{\max}} f_0(\Delta_0) \, d\Delta_0 = 1 - \rho_n(0). \tag{5}$$

The distribution is assumed wide, i. e., $f_0(\Delta_0)$ is non-zero in every point of the interval. In principle, $f_0(\Delta_0)$ can be random or not, but the former case seems more frequently occurring.

At $T \neq 0$ the superconducting components with $T_{c0} < T$, i.e., with $\Delta_0 < \Delta^*(T) = \frac{\pi}{\gamma} T$, lose their superconducting properties. The normal fraction of the current carriers in the sample is

$$\rho_n(T) = \rho_n(0) + \int_0^{\Delta^*(T)} f_0(\Delta_0) \, d\Delta_0, \tag{6}$$

whereas the remaining superconducting part is

$$\rho_s(T) = \int_{\Delta^*(T)}^{\Delta_0^{\max}} f_0(\Delta_0) \, d\Delta_0. \tag{7}$$

Due to the condition (B), the components, possessing at $T = 0$ OP's within the interval $[\Delta_0, \Delta_0 + d\Delta_0]$, at $T \neq 0$ acquire OP's within the interval $[\Delta, \Delta + d\Delta]$, where $\Delta = \Delta_{\mathrm{BCS}}(\Delta_0, T)$. This conversion is expressed by an equation

$$f(\Delta, T) \, d\Delta = f_0(\Delta_0) \, d\Delta_0. \tag{8}$$

Here $f(\Delta, T)$ is a function characterizing a new distribution of components in the interval $0 < \Delta < \Delta^{\max}(T)$, where $\Delta^{\max}(T) = \Delta_{\mathrm{BCS}}(\Delta_0^{\max}, T)$. This equation is a consequence of (i) the supposed domain structure permanence, (ii) the constant current carrier density, and (iii) the independence between superconducting components. Then, the function $\rho_s(T)$ takes the form

$$\rho_s(T) = \int_0^{\Delta^{\max}(T)} f(\Delta, T) \, d\Delta. \tag{9}$$

As for any investigated property P, each component, being superconducting or not, makes its contribution to the measured (averaged) value $\langle P \rangle$:

$$\langle P(T) \rangle = P_n(T)\rho_n(T) + \int_0^{\Delta^{\max}(T)} P_s(\Delta, T) \, f(\Delta, T) \, d\Delta. \tag{10}$$

This formula is valid (with restriction given above) for additive quantities, such as, e.g., the specific heat. But what about, for example, the penetration depth λ_L? Really, in the situation when the superconducting gap changes (and in fact goes to zero) on a very short length scale, even the notion of the penetration depth becomes questionable. Moreover, since each of our elementary volumes includes an ensemble of domains with *different* parameters $\lambda_{L,i}$'s, the matter becomes much more entangled. Nevertheless, even in this situation one may introduce an effective penetration depth λ_L^{eff} and measure its T-dependence. Really, the measured electromagnetic response of the nonhomogeneous superconductor is the *sum* of individual domain responses from the sample surface layer. The quantity λ_L^{eff} is a parameter that is extracted from the essentially averaged experimental data treated as obtained for a homogeneous BCS superconductor. In the specific case of cuprates the domain sizes d_i are substantially smaller than the intrinsic penetration depths $\lambda_{L,i}$ for each domain and, therefore, the effective λ_L^{eff}. We conceive that within such a context the calculation of $\lambda_L^{\mathrm{eff}}(T)$ as a weighted quantity is at least qualitatively reasonable.

The first term in Eq. (10) describes the contribution $\langle P(T)\rangle_n$ of the normal fraction. It is well-known and will not be considered below. The last term corresponds to the contribution $\langle P(T)\rangle_s$ of the superconducting electrons (holes). Since $f(\Delta = 0, T) = f_0[\Delta_0 = \Delta^*(T)] \neq 0$, for each T there is a nonvanishing portion of superconducting components with $\Delta \to 0$. It is their contribution that leads to the deviation of the temperature behavior $\langle P(T)\rangle_s$ from the classical one. To make our statement even more sound, we suggest that the low-T asymptotics of the types (1)–(4) hold true for each superconducting component up to the relevant critical temperature T_{c0}. The allowance for the exact dependences may only strengthen our standpoint.

Note that the low-T expressions for the properties $P_s(\Delta, T)$ under consideration as well as for others in the framework of the BCS scheme depend on T and on Δ_0 rather than on Δ value. Accordingly, due to Eq. (8) the contribution $\langle P(T)\rangle_s$ can be rewritten as follows:

$$\langle P(T)\rangle_s = \int_{\Delta^*(T)}^{\Delta_0^{\max}} P_s(\Delta_0, T) f_0(\Delta_0) \, d\Delta_0. \tag{11}$$

To consider the properties of interest at low T simultaneously, the general form

$$P_s(\Delta_0, T) = A\Delta_0^m T^l \exp\left(-\frac{\Delta_0}{T}\right) \tag{12}$$

of the Eqs. (1)–(4) is used. The distribution function $f_0(\Delta_0)$ can be expanded into the series

$$f_0(\Delta_0) = \frac{1}{\Delta_0^{\max}} \sum_{k=k_0}^{\infty} B_k \left(\frac{\Delta_0}{\Delta_0^{\max}}\right)^k. \tag{13}$$

Substituting Eqs. (12) and (13) into Eq. (11) we obtain

$$\langle P(T)\rangle_s = \frac{AT^{l+m+1}}{\Delta_0^{\max}} \sum_{k=k_0}^{\infty} B_k \left(\frac{T}{\Delta_0^{\max}}\right)^k \int_{\frac{\Delta^*(T)}{T}}^{\frac{\Delta_0^{\max}}{T}} x^{m+k} e^{-x} \, dx. \tag{14}$$

Within an accuracy of the made approximations and for temperatures $T \ll \Delta_0^{\max}$ we may extend the upper limit of integration to infinity, so

$$\langle P(T)\rangle_s \approx AT^{l+m} \left(\frac{T}{\Delta_0^{\max}}\right) \sum_{k=k_0}^{\infty} B_k \left(\frac{T}{\Delta_0^{\max}}\right)^k \Gamma\left[m+k+1, \frac{\Delta^*(T)}{T}\right],$$

$$\tag{15}$$

where $\Gamma(a, x)$ is the incomplete gamma function [124]. Since $\frac{\Delta^*(T)}{T} = \frac{\pi}{\gamma}$, the apparently dominant exponential dependence of $\langle P(T) \rangle_s$ on $\left(-\frac{1}{T} \right)$ resulting from the second argument of $\Gamma(a, x)$ disappears altogether, whatever the particular value of k_0.

One more important result of this formula is that in the framework of the proposed model the measured properties of the superconducting components $\langle P(T) \rangle_s$ at low temperatures are *insensible* to the particular profile of the distribution function $f_0(\Delta_0)$ at large Δ_0. Hence, for $T \ll \Delta_0^{\max}$ a few first terms of the series (15) constitute a good approximation. Restricting ourselves to the leading k_0-term we obtain

$$\langle P(T) \rangle_s = AB_{k_0}(\Delta_0^{\max})^{l+m}\Gamma\left(m + k_0 + 1, \frac{\pi}{\gamma} \right) \left(\frac{T}{\Delta_0^{\max}} \right)^M, \qquad (16)$$

with $M = k_0 + l + m + 1$. The corrections to this expression are of the next order in $\frac{T}{\Delta_0^{\max}}$. This justifies the validity of substituting the upper limit of the integral in Eq. (14) by infinity. At the same time, this makes eligible the evaluating of the $\langle P(T) \rangle_s$ contribution in Eq. (10) using the low-T asymptotics $P_s(\Delta_0, T)$ in the integrand instead of the exact value $P_s(\Delta, T)$. Indeed, the T-dependences of various parameters in the BCS theory are induced by the T-behavior of the gap Δ [1], e.g., the exponential multipliers in Eqs. (1)–(4) originate from that in the low-T asymptotics of $\Delta(T)$. Since $\Delta(T \to T_c) \propto (T_c - T)^{1/2}$ in the BCS theory, the considered parameters have at $T \to T_c$ the power-like asymptotics as well. Thus, the use of exact functional dependences $P_s(\Delta, T)$ the more so has to result *not* in *exponential* but power-law dependences $\langle P(T) \rangle_s$.

The generalization of the Eq. (11) to other types of pairing symmetry is straightforward. One should take into account that not only the dependence $P_s(\Delta, T)$ will have another functional form but also the limits of the integral will change. It is clear that the T-dependences of P_s and $\langle P_s \rangle$ may be different for any OP symmetry.

One should note that, in each specific experiment only a certain lowest temperature $T_{\lim}$ is accessible, so that, according to the Eq. (11), only gap values down to $\Delta_0^{\lim} = \frac{\pi}{\gamma}T_{\lim}$ are relevant. Hence, the restriction imposed above on the distribution function $f_0(\Delta_0)$ to extend down to $\Delta_0 = 0$ may be weakened. Namely, $f_0(\Delta_0)$ should be nonzero for $\Delta_0 > \Delta_0^{\lim}$. In the case when the domain ensemble possesses the minimal value $\Delta_0^{\min}$ and the lowest accessible $T_{\lim} < \frac{\gamma}{\pi}\Delta_0^{\min}$, the value $\Delta_0^{\min}$ will manifest itself as the exponential factor $\exp\left(-\frac{\Delta_0^{\min}}{T} \right)$ in $\langle P(T) \rangle_s$ (cf. Ref. [1]).

Returning to the Eq. (16), we see that the actual distribution function reveals itself in the final result only through the expansion parameters B_{k_0}

and k_0. Most popular distribution functions [125], namely, normal Gaussian

$$f_G(\Delta_0) = \frac{1}{\Delta_0^{\max}} \sqrt{\frac{2}{\pi \sigma^2}} \left[2\Phi \left(\frac{1}{\sigma} \right) + 1 \right]^{-1} \exp \left[-\frac{1}{2} \left(\frac{\Delta_0 - \Delta_0^{\max}}{\sigma \Delta_0^{\max}} \right)^2 \right], \quad (17)$$

exponential

$$f_E(\Delta_0) = \frac{\sqrt{2}}{\sigma \Delta_0^{\max}} \exp \left(-\frac{\Delta_0 \sqrt{2}}{\sigma \Delta_0^{\max}} \right), \quad (18)$$

and uniform $f_U(\Delta_0)$ ones, where σ is the dispersion parameter normalized to $\Delta_0^{\max}$ and $\Phi(x)$ is the error function [124], have finite values at $\Delta_0 = 0$, so the leading term (16) in the series has the $k_0 = 0$ order of smallness. At the same time, different distribution functions have different values of coefficient B_0. Now it is impossible to make a choice in favor of one of them. The analysis of the heat capacity measurements for various oxides [9, 108] makes us to suggest that the function $f_0(\Delta_0)$ is mainly concentrated in a narrow interval near $\Delta_0 = 0$, which is beneficial for our hypothesis.

Applying the general approach to the Eqs. (1)–(4) and comparing them with Eq. (12) we see that for the specific heat $A = N(0)\sqrt{2\pi}, l = -\frac{3}{2}, m = \frac{5}{2}$ so that for the chosen case $k_0 = 0$ we have $M = 2$ and according to Eq. (16)

$$\langle C_s \rangle \approx B_0 \sqrt{2\pi} N(0) \Delta_0^{\max} \Gamma \left(\frac{7}{2}, \frac{\pi}{\gamma} \right) \left(\frac{T}{\Delta_0^{\max}} \right)^2; \quad (19)$$

for the penetration depth $A = \sqrt{\frac{\pi}{2}}, l = -\frac{1}{2}, m = \frac{1}{2}, M = 1$, and

$$\langle \delta \lambda_L \rangle \approx B_0 \sqrt{\frac{\pi}{2}} \Gamma \left(\frac{3}{2}, \frac{\pi}{\gamma} \right) \frac{T}{\Delta_0^{\max}}; \quad (20)$$

for the thermal conductivity $A = \frac{2n_e \tau_{tr}}{m_e}, l = -1, m = 2, M = 2$, and

$$\langle \kappa_e \rangle \approx B_0 \frac{2n_e \tau_{tr}}{m_e} \Delta_0^{\max} \Gamma \left(3, \frac{\pi}{\gamma} \right) \left(\frac{T}{\Delta_0^{\max}} \right)^2; \quad (21)$$

for the ultrasound attenuation $A = 2, l = 0, m = 0, M = 1$, and

$$\langle \delta \alpha \rangle \approx 2B_0 \Gamma \left(1, \frac{\pi}{\gamma} \right) \frac{T}{\Delta_0^{\max}}. \quad (22)$$

These results correlate well with experimental data (see the next Section). For other possible distribution functions with $k_0 > 0$ the preceding results will remain power-law, although with larger M. In 2D-superconductors, such as cuprates, the value $k_0 = 0$ corresponds to linear objects, i.e., lines or edges of normal regions, consisting of "nodes" ($\Delta_0 = 0$) in the real space. Point-like zeros would lead to $k_0 = 1$, so that the relevant power-law exponents would increase by one.

From the methodological point of view it is of interest to indicate an analogy between our approach dealing with the Δ-distribution in the real space and the Abrikosov's introduction [126] of the distribution function $P(\Delta)$ for the OP Δ anisotropic in the momentum space, with the anisotropy being quite general and including both d-wave and extended s-wave symmetries.

4. Discussion

The coherence length ξ for the electron-doped NCCO was shown [106] to exceed substantially that for the hole-doped YBCO (both quoted in Sec. 2). Moreover, making also allowance for that superconductivity of NCCO exists in the narrow range $0.14 < x < 0.15$ and $y \leq 0.01$ [127], it is natural to conclude that the spread of Δ assumed in our model is not large enough to validate the averaging procedure. Thus, NCCO should manifest its intrinsic exponential low-T asymptotics which is indeed the case [106].

On the other hand, tunnel spectra of YBCO show a large spread of Δ magnitudes [128] which is favorable for our interpretation. The growth with x of structural domains with different nonoptimal (for a nominal stoichiometry) Δ's and the attended widening of the distribution function $f(\Delta)$ may explain the increase of the numerical factor in the observed linear-T term of $\delta\lambda_L(T)$ for YBCO [129].

Comparing our theory with experiment, we may state that it describes properly the respective power-law exponents of various properties for oxides as well as the specific heat asymptotics for Nb$_3$Sn. At the same time, it can account for substantial contributions to the low-T asymptotics for cuprates. To show this, let us compare our results for $\langle\delta\lambda_L\rangle$ with those of the pure d-wave approach. The choice of this quantity was made because it usually applied as a sensitive probe to distinguish between different pairing mechanisms. For the dispersion test value $\sigma = 1$ we obtain $B_0 \approx 0.29$ for the distribution function $f_G(\Delta_0)$, $B_0 \approx 1.41$ for $f_E(\Delta)$, and $B_0 = 1$ for $f_U(\Delta)$. The corresponding values 0.10, 0.50, and 0.35 of the coefficient before the term $\frac{T}{\Delta_0^{\max}}$ in Eq. (20) correlate well with the pure d-based calculation result $\delta\lambda_L \approx \frac{T}{\Delta_0}\ln 2$ [16], which is consistent with the experiments for YBCO [75]. The observations of different exponents for $\delta\lambda_L(T)$ in various samples [15, 75, 76, 77, 78, 79] may reflect their dissimilar nonhomogeneous structures, leading to a change-over from one $f_0(\Delta_0)$ to another with different k_0's.

Note that there is also another approach [130], valid both for s- and d-OP symmetry, which is based on the proximity effect in the S-N layer structures of cuprates and fits the experimental data on $\lambda_L(T)$. A possibility of the transformation of the dependence (2) into the power-law one with $M \leq 1$ due to the proximity effect was demonstrated in Ref. [131] for Nb/Al bilayer films.

One should pay attention that according to the proposed theory the power exponents M's in the $\langle P(T) \rangle_s$ dependences, both thermal and transport, are interrelated through the common distribution function $f_0(\Delta_0)$. It makes our hypothesis about the influence of the latter on the experimental dependence $\langle P(T) \rangle_s$ checkable. Moreover, if our assumption concerning the role of the mesoscopic domain structure is valid, then going beyond the low-T interval makes it possible to determine the actual $f_0(\Delta_0)$ dependence. The structure of the integral equations (10) or (11) permits to fulfill this procedure unequivocally. Of course, our theory does not pretend to explain the specific form of $f_0(\Delta_0)$ derived in such a manner. To solve the arising problems a microscopical theory should be invoked.

In connection with this we note that the alternative feasibility of *different* kinds of asymptotics for thermodynamic quantities on the one hand and various transport properties on the other provided the *same* distribution function should not be overlooked. Really, the latter may possess their specific scales interfering into the averaging procedure. If such a scale is bigger than the domain size d_i, the type of averaging suggested above is no longer meaningful. Instead, a relevant process characterized by an average Δ_0^{av} from the interval $\{\Delta_0^{min}, \Delta_0^{max}\}$, would maintain the primordial exponential BCS-like dependence. It should be emphasized that our speculations concerning the penetration depth in Sec. 3 are not applicable in this case. The robustness of the exponential dependence may arise, e.g., for the nuclear spin-lattice relaxation rate, so that the contact interaction of the magnetic moment with itinerant electrons [132] would be modified. Unfortunately, we do not know at present the origin of the possible corresponding length but, according the observed deviations from the linear-T plot [133, 134], it should be T-dependent. It is not improbable that superconducting fluctuations, anomalously large for cuprates [134, 135, 136], are essential here. If such an *ad hoc* hypothesis is true, it is possible to reconcile the dependence $C_s \sim T^2$ [23] with the exponential drop of T_1^{-1} for $T \ll T_c$ and the existence of the Hebel-Slichter peak near T_c [137] for the Laves phase compound $CeCo_2$, where the mesoscopic nonhomogeneities, required by our theory, may have a structural nature [138].

To summarize, the characteristics of the conventional s-wave superconductor statistically averaged over the spatial distribution of Δ may mimic the d-behavior very plausibly. But the power-law low-T asymptotics obtained here in the framework of the s-wave picture by no means rule out the existence of d-wave superconductivity in high-T_c oxides. The main goal of this article is to stress that although recent experiments demonstrate the mixed $(s+d)$-OP there, the s-wave component inherent at least to the majority of cuprates may be hidden in low-T experiments due to the non-homogeneous structure of the samples.

Acknowledgements

We are grateful to all colleagues who sent us the reprints of their works and especially to V. V. Kabanov, R. A. Klemm, M. A. Lorenz, D. Mihailovic, O. V. Misochko, M. Mößle, L. Ozyuzer, C. Panagopoulos, and Ya. G. Ponomarev for providing us with the reprints prior to publication. We are also grateful to N. H. Andersen, J. F. Annett, M. Ausloos, T. Ekino, K. Maki, and D. Pavuna for illuminating discussions. This work was supported, in part, by the Ukrainian State Foundation for Fundamental Researches (Grant 2.4/100).

References

1. Abrikosov, A. A. (1987) *Fundamentals of the Theory of Metals*, North-Holland, Amsterdam.
2. (1977) V. L. Ginzburg and D. A. Kirzhnitz (eds.), *Problem of High-Temperature Superconductivity*, Nauka, Moscow, in Russian.
3. Vonsovskii, S. V., Izyumov, Yu. A., and Kurmaev, E. Z. (1977) *Superconductivity of Transition Metals, Their Alloys and Compounds*, Nauka, Moscow, in Russian.
4. Hulm, J. K., Ashkin, M., Deis, D. W., and Jones, C. K. (1970) *Progr. Low Temp. Phys.* **6**, 205.
5. Allen, P. B. and Mitrović, B. (1982) *Solid State Phys.* **37**, 1.
6. Geilikman, B. T. (1979) *Studies in Low Temperature Physics*, Atomizdat, Moscow, in Russian.
7. (1984) T. Matsubara and A. Kotani (eds.), *Superconductivity in Magnetic and Exotic Materials*, Springer Verlag, Berlin.
8. Brandow, B. (1998) *Phys. Rep.* **296**, 1.
9. Gabovich, A. M. and Moiseev, D. P. (1986) *Usp. Fiz. Nauk* **150**, 599 [(1986) *Sov. Phys. Usp.* **29**, 1135].
10. Maple, M. B. (1998) *J. Magn. Magn. Mat.* **177-181**, 18.
11. Fisk, Z. and Sarrao, J. L. (1997) *Annu. Rev. Mat. Sci.* **27**, 35.
12. Hauck, J. and Mika, K. (1998) *Supercond. Sci. Technol.* **11**, 614.
13. Ginzburg, V. L. and Maksimov, E. G. (1992) *Sverkhprovodimost* **5**, 1543.
14. Gabovich, A. M. and Voitenko, A. I. (1996) *Physica C* **258**, 236.
15. Annett, J. F., Goldenfeld, N. D., and Leggett, A. J. (1996) in D. M. Ginsberg (ed.), *Physical Properties of High Temperature Superconductors V*, World Scientific, River Ridge, N. J., p. 375.
16. Scalapino, D. J. (1995) *Phys. Rep.* **250**, 329.
17. Ausloos, M. and Houssa, M. (1999) *Supercond. Sci. Technol.* **12**, R1.
18. Annett, J. F. (1999) *Physica C* **317-318**, 1.
19. Timusk, T. (1999) *Physica C* **317-318**, 18.
20. Won, H. and Maki, K. (1999) in M. Ausloos and S. Kruchinin (eds.), *Symmetry and Pairing in Superconductors*, Kluwer Academic, Dordrecht, p. 3.
21. Sigrist, M., Agterberg, D., Furusaki, A., Honerkamp, C., Ng, K. K., Rice, T. M., and Zhitomirsky, M. E. (1999) *Physica C* **317-318**, 134.

22. Sugawara, H., Inoue, O., Kobayashi, Y., Sato, H., Nishigaki, T., Aoki, Y., Sato, H., Settai, R., and Onuki, Y. (1995) *J. Phys. Soc. Jpn.* **64**, 3639.
23. Aoki, Y., Nishigaki, T., Sugawara, H., and Sato, H. (1997) *Phys. Rev.* **B55**, 2768.
24. Wosnitza, J. (1999) *Physica C* **317-318**, 98.
25. Levin, K., Kim, J. H., Lu, J. P., and Si, Q. (1991) *Physica C* **175**, 449.
26. Pruschke, Th., Jarrell, M., and Freericks, J. K. (1995) *Adv. Phys.* **44**, 187.
27. Berger, H., Forró, L., and Pavuna, D. (1998) *Europhys. Lett.* **41**, 531.
28. Liang, W. Y. (1998) *J. Phys.: Condens. Matter* **10**, 11365.
29. Tsuei, C. C., Kirtley, J. R., Ren, Z. F., Wang, J. H., Raffy, H., and Li, Z. Z. (1997) *Nature* **387**, 481.
30. Kouznetsov, K. A., Sun, A. G., Chen, B., Katz, A. S., Bahcall, S. R., Clarke, J., Dynes, R. C., Gajewski, D. A., Han, S. H., Maple, M. B., Giapintzakis, J., Kim, J.-T., and Ginsberg, D. M. (1997) *Phys. Rev. Lett.* **79**, 3050.
31. Sergeeva, G. G., Stepanovskii, Yu. P., and Chechkin, A. V. (1998) *Fiz. Nizk. Temp.* **24**, 1029 [(1998) *Low Temp. Phys.* **24**, 771].
32. Klemm, R. A. (1998) in S. M. Bose and K. B. Garg (eds.), *High Temperature Superconductivity: Ten Years after Discovery*, Narosa Publishing House, New Delhi, India, p. 179.
33. Klemm, R. A., Rieck, C. T., and Scharnberg, K. (1998) in I. Bozović and D. Pavuna (eds.), *Superconducting Superlattices II: Native, Artificial*, SPIE, Vol. 3480, p. 209.
34. Klemm, R. A. (1998) *Int. J. Mod. Phys. B* **12** , 2920.
35. Bhattacharya, A., Žutić, I., Valls, O. T., Goldman, A. M., Welp, U., and Veal, B. (1999) *Phys. Rev. Lett.* **82**, 3132.
36. Bhattacharya, A., Goldman, A. M., Žutić, I., Valls, O. T., Welp, U., and Veal, B. (1999) *J. Supercond.* **12** , 99.
37. Tsay, Y. N., Li, Q., Zhu, Y., Suenaga, M., Gu, G. D., and Koshizuka, N. (1998) in I. Bozović and D. Pavuna (eds.), *Superconducting Superlattices II: Native, Artificial*, SPIE, Vol. 3480, p. 21.
38. Gabovich, A. M. and Voitenko, A. I. (1999) in M. Ausloos and S. Kruchinin (eds.), *Symmetry and Pairing in Superconductors* , Kluwer Academic, Dordrecht, p. 187.
39. Gabovich, A. M. and Voitenko, A. I. (1997) *Phys. Rev. B* **55**, 1081.
40. Gupta, A. K. and Ng, K.-W. (1998) *Phys. Rev. B* **58**, 8901.
41. Ekino, T., Sezaki, Y., and Fujii, H. (1999) *Phys. Rev. B* **60**, 6916.
42. Ponomarev, Ya. G., Shabalin, M. E., Kuzmich, A. I., Uk, K. K., Sudakova, M. V., Chesnokov, S. N., Aminov, B. A., Lorenz, M., Müller, G., Piel, H., and Hein, M. (1998) in *Abstracts of the XXXI Workshop on Low Temperature Physics, Moscow, 2-3 December 1998*, Moscow State Univ., Moscow, p. 228, in Russian.
43. Ponomarev, Ya. G., Tsokur, E. B., Sudakova, M. V., Tchesnokov, S. N., Shabalin, M. E., Lorenz, M. A., Hein, M. A., Müller, G., Piel, H., and Aminov, B. A. (1999) *Solid State Commun.* **111**, 513.
44. Barone, A. and Paterno, G. (1982) *The Physics and Applications of the Josephson Effect*, John Wiley and Sons, New York.
45. Barash, Yu. S. and Svidzinskii, A. A. (1997) *Zh. Eksp. Teor. Fiz.* **111**, 1120 [(1997) *Sov. Phys. JETP* **84**, 619].
46. Kabanov, V. V., Demsar, J., Podobnik, B., and Mihailovic, D. (1999) *Phys. Rev. B* **59**, 1497.
47. Demsar, J., Podobnik, B., Kabanov, V. V., Wolf, Th., and Mihailovic, D. (1999) *Phys. Rev. Lett.* **82**, 4918.
48. Timusk, T. and Statt, B. (1999) *Rep. Prog. Phys.* **62**, 61.
49. Gabovich, A. M. and Voitenko, A. I. (1997) *J. Phys.: Condens. Matter* **9**, 3901.
50. Ponomarev, Ya. G., Aminov, B. A., Brandt, N. B., Hein, M., Khi, C. S., Kresin, V. Z., Müller, G., Piel, H., Rosner, K., Tchesnokov, S. V., Tsokur, E. B., Wehler, D., Winzer, R., Wolfe, Th., Yarygin, A. V., and Yusupov, K. T. (1995) *Phys. Rev.*

B **52**, 1352.

51. Ponomarev, Ya. G., Khi, C. S., Uk, K. K., Sudakova, M. V., Tchesnokov, S. N., Lorenz, M. A., Hein, M. A., Müller, G., Piel, H., Aminov, B. A., Krapf, A., and Kraak, W. (1999) *Physica C* **315**, 85.
52. Lake, B., Aeppli, G., Mason, T. E., Schröder, A., McMorrow, D. F., Lefmann, K., Isshiki, M., Nohara, M., Takagi, H., and Hayden, S. M. (1999) *Nature* **400**, 43.
53. Sun, A. G., Gajewski, D. A., Maple, M. B., and Dynes, R. C. (1994) *Phys. Rev. Lett.* **72**, 2267.
54. Katz, A. S., Sun, A. G., and Dynes, R. C. (1995) *Appl. Phys. Lett.* **66**, 105.
55. Mößle, M. and Kleiner, R. (1999) *Phys. Rev. B* **59**, 4486.
56. Srikanth, H., Willemsen, B. A., Jacobs, T., Sridhar, S., Erb, A., Walker, E., and Flükiger, R. (1997) *Phys. Rev. B* **55**, 14733.
57. Misochko, O. V. and Uchida, S. (1998) *Phys. Lett. A* **248**, 423.
58. Misochko, O. V. and Gu, G. (1999) *Phys. Rev. B* **59**, 11183.
59. Misochko, O. V., Sakai, K., Nakashima, S., and Gu, G. (1999) *Phys. Rev. B* **60**, 1326.
60. Wu, W. C. and Carbotte, J. P. (1998) *Phys. Rev. B* **57**, 5614.
61. Xiang, T., Panagopoulos, C., and Cooper, J. R. (1998) *Int. J. Mod. Phys. B* **12**, 1007.
62. Phillips, N. E., Buffeteau, B., Calemczuk, R., Dennis, K. W., Emerson, J. P., Fisher, R. A., Gordon, J. E., Hargreaves, T. E., Marcenat, C., McCallum, R. W., O'Connor, A. S., Schilling, A., Woodfield, B. F., and Wright, D. A. (1999) *J. Supercond.* **12**, 105.
63. Kaldis, E. (1997) in E. Kaldis, E. Liarokapis, and K. A. Müller (eds.), *High-T_c Superconductivity 1996: Ten Years after the Discovery*, Kluwer Academic, Dordrecht, p. 411.
64. Jorgensen, J. D. (June 1991) *Phys. Today* **44**, 34.
65. Egami, T. and Billinge, S. J. L. (1994) *Prog. Mat. Sci.* **38**, 359.
66. Egami, T. and Billinge, S. J. L. (1996) in D. M. Ginsberg (ed.), *Physical Properties of High Temperature Superconductors V*, World Scientific, River Ridge, N. J., p. 265.
67. Kirtley, J. R., Mota, A. C., Sigrist, M., and Rice, T. M. (1998) *J. Phys.: Condens. Matter* **10**, L97.
68. Kostić, P., Veal, B., Paulikas, A. P., Welp, U., Todt, V. R., Gu, C., Geiser, U., Williams, J. M., Carlson, K. D., and Klemm, R. A. (1997) *Phys. Rev. B* **55**, 14649.
69. Geim, A. K., Dubonos, S. V., Lok, J. G. S., Henini, M., and Maan, J. C. (1998) *Nature* **396**, 144.
70. Vieland, L. J. and Wicklund, A. W. (1968) *Phys. Rev.* **166**, 424.
71. Brock, J. C. F. (1969) *Solid State Commun.* **7**, 1789.
72. Stewart, G. R. and Brandt, B. L. (1984) *Phys. Rev. B* **29**, 3908.
73. Moler, K. A., Sisson, D. L., Urbach, J. S., Beasley, M. R., Kapitulnik, A., Baar, D. J., Liang, R., and Hardy, W. N. (1997) *Phys. Rev. B* **55**, 3954.
74. Monien, H., Scharnberg, K., Tewordt, L., and Walker, D. (1987) *Solid State Commun.* **61**, 581.
75. Hardy, W. N., Bonn, D. A., Morgan, D. C., Liang, R., and Zhang, K. (1993) *Phys. Rev. Lett.* **70**, 3999.
76. Ma, Z., Taber, R. C., Lombardo, L. W., Kapitulnik, A., Beasley, M. R., Merchant, P., Eom, C. B., Hou, S. Y., and Phillips, J. M. (1993) *Phys. Rev. Lett.* **71**, 781.
77. Lee, J. Y., Paget, K. M., Lemberger, T. R., Foltyn, S. R., and Wu, X. (1994) *Phys. Rev. B* **50**, 3337.
78. Bonn, D. A., Kamal, S., Zhang, K., Liang, R., Baar, D. J., Klein, E., and Hardy, W. N. (1994) *Phys. Rev. B* **50**, 4051.
79. Walter, H., Prusseit, W., Semerad, R., Kinder, H., Assmann, W., Huber, H., Burkhardt, H., Rainer, D., and Sauls, J. A. (1998) *Phys. Rev. Lett.* **80**, 3598.
80. Klein, N., Tellmann, N., Schulz, H., Urban, K., Wolf, S. A., and Kresin, V. Z.

(1993) *Phys. Rev. Lett.* **71**, 3355.

81. Panagopoulos, C., Cooper, J. R., Xiang, T., Peacock, G. B., Gameson, I., and Edwards, P. P. (1997) *Phys. Rev. Lett.* **79**, 2320.

82. Panagopoulos, C., Cooper, J. R., Athanassopoulou, N., and Chrosch, J. (1996) *Phys. Rev. B* **54**, 12721.

83. Panagopoulos, C., Tallon, J. L., and Xiang, T. (1999) *Phys. Rev. B* **59**, 6635.

84. Fournier, P., Mahanty, P., Maiser, E., Darzens, S., Venkatesan, T., Lobb, C. J., Czjzek, G., Webb, R. A., and Greene, R. L. (1998) *Phys. Rev. Lett.* **81**, 4720.

85. Graf, M. J., Yip, S-K., Sauls, J. A., and Rainer, D. (1996) *Phys. Rev. B* **53**, 15147.

86. Houssa, M. and Ausloos, M. (1997) *Phys. Rev. B* **56**, 953.

87. Taillefer, L., Lussier, B., Gagnon, R., Behnia, K., and Aubin, H. (1997) *Phys. Rev. Lett.* **79**, 483.

88. Movshovich, R., Hubbard, M. A., Salamon, M. B., Balatsky, A. V., Yoshizaki, R., Sarrao, J. L., and Jaime, M. (1998) *Phys. Rev. Lett.* **80**, 1968.

89. Bhattacharya, S., Higgins, M. J., Johnston, D. C., Jacobson, A. J., Stokes, J. P., Lewandowski, J. T., and Goshorn, D. P. (1988) *Phys. Rev. B* **37**, 5901.

90. Xu, M.-F., Baum, H.-P., Schenstrom, A., Sarma, B. K., Levy, M., Sun, K. J., Toth, L. E., Wolf, S. A., and Gubser, D. U. (1988) *Phys. Rev. B* **37**, 3675.

91. Markiewicz, R. S. (1997) *J. Phys. Chem. Sol.* **58**, 1179.

92. Božin, E. S., Billinge, S. J. L., Kwei, G. H., and Takagi, H. (1999) *Phys. Rev. B* **59**, 4445.

93. Sleight, A. W. (June 1991) *Phys. Today* **44**, 24.

94. Yanovskii, V. K., Voronkova, V. I., and Vodolazskaya, I. V. (1991) in *Physical Properties of High-T_c Superconductors*, Basis, Moscow, p. 3, in Russian.

95. Cannelli, G., Cantelli, R., Cordero, F., and Trequattrini, F. (1992) *Supercond. Sci. Technol.* **5**, 247.

96. Andersen, N. H., von Zimmermann, M., Frello, T., Käll, M., Mønster, D., Lindgård, P.-A., Madsen, J., Niemöller, T., Poulsen, H. F., Schmidt, O., Schneider, J. R., Wolf, Th., Dosanjh, P., Liang, R., and Hardy, W. N. (1999) *Physica C* **317-318**, 259.

97. Nagaev, E. L. (1995) *Usp. Fiz. Nauk* **165**, 529 [(1996) *Physics Usp.* **48**, 997].

98. Krivoglaz, M. A. (1984) *Diffuse Scattering of X-rays and Neutrons by Fluctuation Inhomogeneities in Nonideal Crystals*, Naukova Dumka, Kiev, in Russian.

99. Crawford, M. K., Harlow, R. L., McCarron, E. M., Tozer, S. W., Huang, Q., Cox, D. E., and Zhu, Q. (1997) in E. Kaldis, E. Liarokapis, and K. A. Müller (eds.), *High-T_c Superconductivity 1996: Ten Years after the Discovery*, Kluwer Academic, Dordrecht, p. 281.

100. Annett, J. F. and Goldenfeld, N. (1992) *J. Low Temp. Phys.* **89**, 197.

101. Mesot, J. and Furrer, A. (1997) *J. Supercond.* **10**, 623.

102. Mesot, J. and Furrer, A. (1998) in A. Furrer (ed.), *Neutron Scattering in Layered Copper-Oxide Superconductors*, Kluwer Academic, Dordrecht, p. 335.

103. Iliev, M. N. (1999) in E. Faulques (ed.), *Spectroscopy of Superconducting Materials*, Amer. Chem. Soc., Washington.

104. Bill, A., Hizhnyakov, V., Nevedrov, D., Seibold, G., and Sigmund, E. (1997) *Z. Phys. B* **104**, 753.

105. Hammel, P. C. and Scalapino, D. J. (1996) *Phil. Mag. B* **74**, 523.

106. Anlage, S. M., Wu, D.-H., Mao, J., Mao, S. N., Xi, X. X., Venkatesan, T., Peng, J. L., and Greene, R. L. (1994) *Phys. Rev.* **B50**, 523.

107. Methfessel, C. E., Stewart, A. R., Matthias, B. T., and Patel, C. K. N. (1980) *Proc. Nat. Acad. Sci. USA* **77**, 6307.

108. Dikin, D. A., Dmitriev, V. M., Isakina, A. P., and Prokhvatilova, A. I. (1990) *Fiz. Nizk. Temp.* **16**, 635.

109. Dooglav, A. V., Alloul, H., Bakharev, O. N., Berthier, C., Egorov, A. V., Horvatic, M., Krjukov, E. V., Mendels, P., Sakharov, Yu. A., and Teplov, M. A. (1998) *Phys. Rev. B* **57**, 11792.

212

110. Veal, B. W., Paulikas, A. P., You, H., Shi, H., Fang, Y., and Downey, J. W. (1990) *Phys. Rev. B* **42**, 6305.
111. Fietz, W. H., Quenzel, R., Ludwig, H. A., Grube, K., Schlachter, S. I., Hornung, F. W., Wolf, T., Erb, A., Kläser, M., and Müller-Vogt, G. (1996) *Physica C* **270**, 258.
112. Sieburger, R. and Schilling, J. S. (1991) *Physica C* **173**, 423.
113. Sadewasser, S., Schilling, J. S., Wagner, J. L., Chmaissem, O., Jorgensen, J. D., Hinks, D. G., and Dabrowski, B. (1999) *Phys. Rev. B* **60**, 9827.
114. Weger, M. and Goldberg, I. (1973) *Solid State Phys.* **28**, 2.
115. Golovashkin, A. I. and Lykov, A. N. (1988) *Trudy FIAN* **190**, 144.
116. Shen, L. Y. L. (1972) *Phys. Rev. Lett.* **29** , 1082.
117. Rowell, J. M. and Schmidt, P. H. (1976) *Appl. Phys. Lett.* **29**, 622.
118. Moore, D. F., Rowell, J. M., and Beasley, M. R. (1976) *Solid State Commun.* **20**, 305.
119. Vedeneev, S. I. (1983) *Trudy FIAN* **148**, 47.
120. Vedeneev, S. I., Golovashkin, A. I., Levchenko, I. S., and Motulevich, G. P. (1972) *Zh. Eksp. Teor. Fiz.* **63**, 1010.
121. Oppermann, R. (1988) *Z. Phys. B* **70**, 49.
122. Ziegler, K. (1988) *Commun. Math. Phys.* **120**, 117.
123. Alexandrov, A. S. and Mott, N. F. (1994) *Rep. Prog. Phys.* **57**, 1197.
124. (1964) M. Abramowitz and I. A. Stegun (eds.), *Handbook of Mathematical Functions with Formulas, Graphs and Mathematical Tables*, National Bureau of Standards, New York.
125. Feller, W. (1952) *An Introduction to Probability Theory and Its Applications*, John Wiley and Sons, New York, Vol. 1.
126. Abrikosov, A. A. (1993) *Physica C* **214**, 107.
127. Ignatov, A. Yu., Ivanov, A. A., Menushenkov, A. P., Iacobucci, S., and Lagarde, P. (1998) *Phys. Rev. B* **57**, 8671.
128. Ozyuzer, L., Zasadzinski, J. F., Kendziora, C., and Gray, K. E. (1999) *Phys. Rev. B* , to be published.
129. Panagopoulos, C., Cooper, J. R., and Xiang, T. (1998) *Phys. Rev. B* **57**, 13422.
130. Klemm, R. A. and Liu, S. H. (1995) *Phys. Rev. Lett.* **74**, 2343.
131. Pambianchi, M. S., Mao, S. N., and Anlage, S. M. (1995) *Phys. Rev. B* **52**, 4477.
132. Winter, J. (1971) *Magnetic Resonance in Metals* , Clarendon Press, Oxford.
133. Williams, G. V. M., Tallon, J. L., and Loram, J. W. (1998) *Phys. Rev. B* **58**, 15053.
134. Rigamonti, A., Borsa, F., and Carretta, P. (1998) *Rep. Prog. Phys.* **61**, 1367.
135. Randeria, M. and Varlamov, A. A. (1994) *Phys. Rev. B* **50**, 10401.
136. Varlamov, A. A. and Ausloos, M. (1997) in M. Ausloos and A. A. Varlamov (eds.), *Fluctuation Phenomena in High Temperature Superconductors*, Kluwer Academic, Dordrecht, p. 3.
137. Ishida, K., Mukuda, H., Kitaoka, Y., Asayama, K., Sugawara, H., Aoki, Y., and Sato, H. (1997) *Physica B* **237-238**, 304.
138. Pan, V. M., Prokhorov, V. G., and Shpigel, A. S. (1984) *Metal Physics of Superconductors*, Naukova Dumka, Kiev, in Russian.

FIELD DEPENDENT EXPONENTS OF SUPERCONDUCTING FLUCTUATIONS.

M. AUSLOOS
SUPRAS and GRASP,
Sart Tilman, B5
University of Liège, B-4000 Liège, Belgium

AND

M. PEKALA
Department of Chemistry, University of Warsaw,
Al. Zwirki i Wigury 101,
PL-02-089 Warsaw, Poland

Abstract.

The electrical resistivity of a quasi monocrystalline $Bi_2Sr_2CaCu_2O_y$ sample has been re-examined between the superconductivity percolation temperature and the onset temperature in presence of magnetic fields. A scaling law with a specific critical exponent is assumed in three regimes. The field dependent exponents in this "normal state" are shown to provide some argument for an $SO(5)$ symmetry of the order parameter.

1. INTRODUCTION

Since there is still much controversy on the basic microscopic mechanism for inducing superconductivity in the cuprates, it is useful to re-examine a basic property, like the electrical resistivity, which contains the signature of amplitude and phase fluctuations of the order parameter, whence should be sensitive to its symmetry [1]. Moreover it is known that the presence of a magnetic field alters the superconductivity properties and leads to different phases in the (B, T) plane. The phase transition lines in the latter are in fact well probed by measurements of the electrical resistivity ρ. We thus have re-made precise measurements of the latter and analyzed the so-called paraconductivity [2] of a quasi single crystalline $Bi_2Sr_2Ca_1Cu_2O_x$ sam-

213

R. Cloots et al. (eds.), Supermaterials, 213–222.

ple. The resistivity has been measured on a restricted temperature scale encompassing the percolation and critical temperature. The measurements have been taken under various, less than 4 T, magnetic field strengths. We have analyzed the data assuming scaling laws for the paraconductivity in different regimes, thus defining critical-like exponents. In fact, we have assumed like others [2, 3] that a scaling law holds not only at the critical temperature, but also at the percolation temperature, and also at intermediate temperatures. We are somewhat more interested in the range *below the zero field critical temperature* than above the latter, very often examined anyway, see e.g. [4-15] for representative studies on Bi-2212 single crystals or not, though there are likely many others. We have observed that *such exponents vary with the magnetic field strength* in fact. This is unusual from a theoretical point of view. The consequences are discussed in terms of the $SO(5)$ theory of superconductivity [16] in high critical temperature superconductors.

2. SAMPLE PREPARATION

Single crystals of the $Bi-2212$ superconductors were prepared by the travelling solvent floating zone technique starting with stoichiometric composition of the powder oxide precursors. More details were already published and described elsewhere in [17]. Large single crystals with typical dimensions of 10 x 3 x 0.3 mm^3 could be separated from the resulting bars and were characterized applying various methods, including X-ray diffraction, Laue transmission, energy dispersive X-ray analysis and electron scanning microscopy. No trace of the $Bi-2223$ phase was detected by X-rays nor by electrical resistivity and thermomagnetic measurements. Wet chemical analysis indicated that the composition of the single crystals was very close to the nominal one. However the microstructure when analyzed by high resolution polarized light microscopy appeared to be that of a large domain, with defects, and platelet pilings rather than a single crystal. The c-axis was found to be perpendicular to the plane of the sample platelets.

3. ELECTRICAL CONDUCTIVITY MEASUREMENTS

Due to the small sample thinness the transport measurements could be made as if in the ab-plane. The electrical resistivity was measured by a DC four probe technique in a closed cycle refrigerator. The temperature was stabilized with an accuracy better than 50 mK. The temperature of the sample was varied by small steps between 0.1 and 0.5 K. Voltages were recorded after a 30 to 200 seconds delay time, allowing for the system to attain its thermal equilibrium. A magnetic field was applied perpendicularly to the ab-plane. The field was applied at low temperature, and was successively raised up to 4.0 T by 1.0 T steps. The sample was kept in a helium atmosphere in order to exclude changes in oxygen content.

The electrical resistivity at 100 K is of the order of 5 $\mu\Omega$m and is comparable to values of 2.0, 4.0, 6.0, and 40.0 $\mu\Omega$m reported in refs. [4], [7], [5] and [15], respectively. The normal state resistivity rises proportionally to the temperature at the mean rate of 2 x 10^{-2} $\mu\Omega$m/K as for "metallic like" materials. For each applied field strength the transition starts at about 95 K but the resistivity decay becomes less abrupt, and the width larger for higher magnetic fields, as indicated on the $B = 4.0$ T [17]. This is observed by calculating the magnetoresistivity $\Delta\rho(T, B)$ determined by

$$\Delta\rho(T, B) = \rho(T, B) - \rho(T, B = 0). \tag{1}$$

It should be negligible in the normal state, above the onset temperature. Indeed deviations start to appear about 10 K above the (zero field) superconducting transition, thereby indicating the fluctuation upper region. However, a significant raise of $\Delta\rho(T, B)$ occurs in a narrow temperature interval below 95 K. Finally the $\Delta\rho(T, B)$ signal vanishes between 65 and 45 K depending on the field strength, i.e. when it increases between 0.5 and 4.0 T [17].

From this data, we have derived the field dependent percolation temperature line $T_p(B)$ on which $\rho(T, B)$ effectively vanishes. From ref. [17], we have obtained the $B_{c2}(T)$ line, as the value at which the transport entropy vanishes. Both lines are shown on Fig. 1. Recall that the slope of B_{c2} is a measure of the mean inverse correlation length ξ

$$(dB_{c2}/dT_c) = -(\Phi_0/2\pi T_c)[< \xi >]^{-2}, \tag{2}$$

where Φ_0 is the flux quantum.

4. DATA ANALYSIS

The data analysis starts from assuming the simplest scaling law [2, 3] for the paraconductivity $\Delta\sigma(T)$, i.e.

$$\Delta\sigma = \sigma - \sigma_n \cong A(T - T_c)^{-\lambda} \tag{3}$$

where σ_n is the normal state contribution to the electrical conductivity σ, λ is a critical exponent, and A a material constant function of the superconductivity correlation length (or the interlayer spacing, depending on the effective dimensionality of the system). The formula is expected to hold above and below the unknown T_c. It is usual to expect that the normal state electrical resistivity $\rho_n = 1/\sigma_n$ is linear with temperature at high

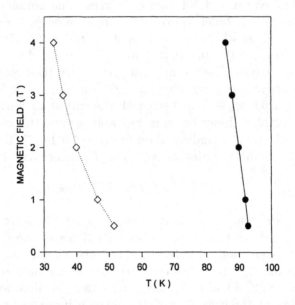

Figure 1. $B_{c2}(T)$ line (•) and field dependent percolation temperature $T_p(B)$ line (◇).

temperature. This is surely verified in the sample for T greater than 200 K as mentioned here above. The inverse of ρ_n is then extrapolated toward low temperatures and substracted from $\sigma = 1/\rho$. Next the logarithmic temperature derivative is numerically obtained, i.e. one defines [13]

$$1/hs1 = -(d/dT)ln(\Delta\sigma). \tag{4}$$

The inverse is obviously

$$hs1 = (1/\lambda)(T - T_c), \tag{5}$$

for T above T_c, while below T_c we write

$$hs1 = (1/\lambda)(T_c - T). \tag{6}$$

This is clearly analogous to a Curie-Weiss law, - an elementary plot allowing to determine the mean field critical temperature. The plot of $hs1$ vs. T shows a marked linear dependence on temperature in several regions where slopes can thus be defined (Fig. 2-4). The temperature regimes depend on the various field strengths. In principle we can adapt the above formulae, in order to define a different exponent λ when T_c is the *mean field (MF) critical temperature* T_{MF}, from Ginzburg-Landau theory, or the *order parameter amplitude critical (c) temperature*, or the *phase coherence*

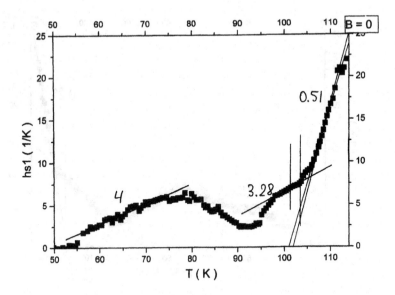

Figure 2. The $hs1$ function, as defined in the text, from measurements of the electrical resistivity of a quasi monocrystalline $Bi_2Sr_2CaCu_2O_y$ sample, at zero external field, as a function of temperature, with "critical exponent" values indicated, i.e. the inverse of the slope(s) assuming a linear data fit in the regions of interest; see Eq. (3)

critical temperature occurring at the percolation (p) temperature T_p in order to examine more precisely the vicinity of each one.

Various fitting and optimization procedures were applied in order to calculate the best parameters of these linear fits, and minimize the error bars before subsequent calculations steps. In so doing the set of exponents λ can be recorded for the various field strengths which we have used, from the positive slopes as shown on Figs. 2-4. The so-called mean field, i.e. arising from superconductivity Gaussian fluctuations λ_G and the so called critical exponent λ_c are shown in Fig. 5, - the latter being divided by a factor of 5 for display purpose.

It is expected that the MF exponent should indeed be equal to 0.5 at $B = 0$, because a theoretical relationship exists between the effective dimensionality of the system D and λ through $\lambda_G = 2 - (D/2)$ [2]. A $D = 3$ value implies $\lambda_G = 0.5$ as observed here. We do not find a value $\lambda_G = 1$ to be expected for highly anisotropic systems, as found in such $Bi-2212$ single crystals [12], i.e. when the coherence length of the fluctuations is shorter than the distance between ab-planes [19]. The value of $\lambda_c(0) = 3.28$ (found for $B = 0$) is rather usual for the behaviour of superconductivity fluctuations percolating in granular materials [20, 21]. This was already found for

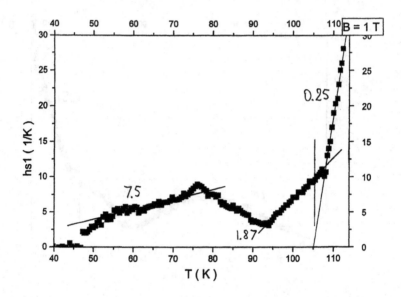

Figure 3. The $hs1$ function, as defined in the text, from measurements of the electrical resistivity of a quasi monocrystalline $Bi_2Sr_2CaCu_2O_y$ sample, for a $B = 1$ T external field, as a function of temperature, with "critical exponent" values indicated, i.e. the inverse of the slope(s) assuming a linear data fit in the regions of interest.

the case of Bi-systems in refs. [8, 22, 23] when it was understood that some (fractal) backbone was the support of superconductivity fluctuations. Let us recall that this regime, away from T_{MF} is a signature of the Josephson coupling between grains, the coupling being directly related to the gap parameter; see e.g. ref. [15]. The $\lambda_c(0)$ value has been shown to be dependent on the measuring current intensity [8] and to vary between 2.9 and 4.3. Finally we show on Fig. 6 the value of the percolation exponent $\lambda_p(0)$, thereby generalizing Eq.(6) to the intergrain electrical path. We find $\lambda_p = 4.0$, of the same order of magnitude as $\lambda_c(0)$. The $\lambda_p(0)$ value also depends on the measuring current intensity [8]. These similar values indicate that the same process is found for the critical regime at which the correlation function of the order parameter *amplitude* diverges, i.e. at T_c, and at the percolation temperature where the correlation function of the order parameter *phase* diverges. In both cases, we find genuine critical fluctuations for $B=0$.

We show in Fig. 5 the behavior of λ_{MF}, $\lambda_c/5$ and $1/\lambda_p$ as a function of the magnetic field. As such the exponents are seen to behave in a smooth but non trivial way on the same scale with a minimum near 1.0 T. We have not found the analytical behaviour of these $\lambda(B)$'s. However we can expect that they saturate toward a finite value for large B. Notice that at high

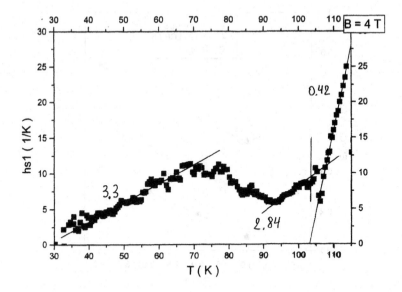

Figure 4. The $hs1$ function, as defined in the text, from measurements of the electrical resistivity of a quasi monocrystalline $Bi_2Sr_2CaCu_2O_y$ sample, for a $B = 4$ T external field, as a function of temperature, with "critical exponent" values indicated, i.e. the inverse of the slope(s) assuming a linear data fit in the regions of interest.

field, say B_{c2}, the superconductivity phenomenon disappears. Moreover at the percolation temperature $(T_c - T_p)/T_c \sim 0.5$. From Fig. 5 it seems that $\lambda_{MF}(B)$, tends to unity for large B. The same holds true for $\lambda_c(B)$ and $1/\lambda_p(B)$. Phenomenological arguments indicate that the value of $1/\lambda_p(B)$ should be of the order of 0.25, if we consider that the conductivity fluctuations are organized on a percolating backbone [24].

5. DISCUSSION

The fact that the exponents of the scaling laws are dependent on an external field is unusual. In basic theories, it is assumed that a critical exponent depends on the effective dimensionality of the system and the order parameter symmetry [24]. In view of these remarks, we can consider that the order parameter symmetry is modified due to the presence of the field. One theory to our knowledge takes such a case into account. In Zhang theory[16, 25], the order parameter has an $SO(5)$ symmetry. This form has been argued to be the condition for the dynamics of the pairing electrons within the Cu-O planes in the framework of a one-band Hubbard model. The model allows to make a connection between antiferromagnetism and

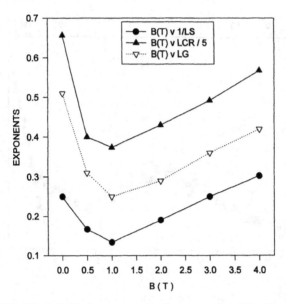

Figure 5. "Modified Critical exponents" of superconductivity fluctuations for a quasi monocrystalline $Bi_2Sr_2CaCu_2O_y$ sample, as a function of the external magnetic field.

superconductivity as a function of oxygen (or hole) doping. The $SO(5)$ symmetry contains as subgroups the $SO(3)$ symmetry of spin rotations and the electromagnetic $SO(2)$ invariance (whose breaking defines the superconductivity phase transition line). The situation is still controversial, and experimental arguments are needed to elucidate the validity of the hypothesis. In Zhang model the order parameter is very sensitive to small effects which break the $SO(5)$ symmetry, like a small magnetic field. It is clear that any finite magnetic field breaks *both* $SO(3)$ **and** $SO(2)$ symmetries. This implies a marked change in the relevant number of components of the order parameter, called the *superspin*, at the critical transition(s) [26]. Thus after a renormalisation group technique and calculation it is necessarily expected that the relevant eigenvalues are field dependent.

6. CONCLUSIONS

In conclusion, the field dependent electrical resistivity of a quasi monocrystalline $Bi_2Sr_2CaCu_2O_y$ sample cotains a set of power law behaviours with specific "critical exponents" in various temperature and magnetic field regimes in the vicinity of the MF and p temperatures, and in between. The exponents are field dependent. This may provide some argument for an $SO(5)$ symmetry of the order parameter.

Acknowledgements

Work supported in part by Polish - Belgian Scientific Exchange Program $UM - 992/16$. We thank NATO for financial aid during this ARW and for making the meeting possible.

References

1. B. Chattopadhyay, D.M. Gaitonde and A. Taraphder, "Fluctuation Effects and Order Parameter Symmetry in the Cuprate Superconductors", *Europhys. Lett.* **34**, 705-710 (1996).

2. A. A. Varlamov and M. Ausloos, "Fluctuation phenomena in superconductors", in *Fluctuation Phenomena in High Temperature Superconductors*, M. Ausloos and A. A. Varlamov, Eds., vol. 32 in the NATO ASI Partnership Sub-Series 3: High Technology (Kluwer, Dordrecht, 1997) p. 3-41.

3. M. Ausloos, S.K. Patapis and P. Clippe, "Superconductivity Fluctuation Effects on Electrical and Thermal Transport Phenomena, $H = 0$, $T > T_{CI}$",in *Physics and Materials Science of High Temperature Superconductors II, Proc. NATO-ASI on High Temperature Superconductors*, Ed. by R. Kossowsky, B. Raveau, D. Wohlleben, and S.K. Patapis, vol. **E209** (Kluwer, Dordrecht, 1992) p.755-785.

4. K. Kadowaki, J.N. Li and J.J.M. Franse, "Superconducting Fluctuation Effects of the Magnetoconductivity in Single-Crystalline $YBa_2Cu_3O_{7-\delta}$ and $Bi_2Sr_2Ca_1Cu_2O_{8+\delta}$", *J. Magnet. Magnet. Mater.* **90-91**, 678-680 (1990).

5. P. Mandal, A. Poddar, A.N. Das, B. Ghosh and P. Choudhury, "Excess conductivity and thermally activated dissipation studies in $Bi_2Sr_2Ca_1Cu_2O_x$ single crystals", *Physica C* **169**, 43-49 (1990).

6. G. Balestrino, M. Marinelli, E. Milani, L. Reggiani, R. Vaglio and A.A. Varlamov, "Excess conductivity in 2:2:1:2-phase Bi-Sr-Ca-Cu-O epitaxail thin films",*Phys. Rev. B* **46**, 14 919-14 921 (1992).

7. M.-O. Mun, S.-I. Lee, S.-H. Suck Salk, H. J. Shin and M.K. Joo, "Conductivity fluctuations in a single crystal of $Bi_2Sr_2Ca_1Cu_2O_x$, *Phys. Rev. B* **48**, 6703-6706 (1993).

8. P. Pureur, R. Menegotto-Costa, P. Rodrigues Jr., J.V. Kunzler, J. Schaf, L. Ghivelder, J.A. Campa and I. Rasines, "Critical and Gaussian Conductivity Fluctuations in $YBa_2Cu_3O_{7-\delta}$ and $Bi_2Sr_2Ca_1Cu_2O_8$", *Physica C* **235-240**, 1939-1940 (1994).

9. A. Ghosh, S.K. Bandyopadhyay, P. Barat, P. Sen and A.N. Basu, "Excess conductivity analysis of α irradiated polycrystalline $Bi - 2212$ superconductor", *Physica C* **255**, 319-323 (1995).

10. A. Pomar, M.V. Ramallo, J. Mosqueira, C. Torron and F. Vidal, "Fluctuation-induced in-plane conductivity, magnetoconductivity, and diamagnetism of $Bi_2Sr_2Ca_1Cu_2O_8$ single crystals in weak magnetic fields", *Phys. Rev. B* **54**, 7470-7480 (1996).

11. G. Heine, W. Lang, X.-L. Wang and X.-Zh. Wang, "Study of Anisotropic Magnetoresistance in the Thermodynamic Fluctuation Regime of Single Crystals of $Bi_2Sr_2CaCu_2O_{8+x}$", *J. Low Temp. Phys.* **10**, 945-955 (1996).

12. R. Menegotto-Costa, P. Pureur, L. Ghivelder, J.A. Campa and I. Rasines, "Gaussian, Three-dimensional-XY, and Lowest Landau-Level Scalings in the Low Field Fluctuation Magnetoconductivity of $Bi_2Sr_2CaCu_2O_8$", *Phys. Rev. B* **56**, 10 836-10 839 (1997).

13. A.R. Jurelo, J.V. Kunzler, J. Schaf, P. Pureur, and J. Rosenblatt, "Fluctuation Conductivity and Microscopic Granularity in Bi-based high temperature superconductors", *Phys. Rev. B* **56**, 14 815-14 821 (1997).

14. C. Boulesteix, Y. Marietti, T. Badèche, H. Tatarenko-Zapolsky, V. Grachev, O. Monnereau, H. Faqir and G. Vacquier, "Use of out-of-plane resistivity for determination of the 2D or 3D character of superconductive fluctuations for Bi-2212 crystals in the mixed state and origin of anomalous high out-of-plane resistivity for weakly oxidized crystals", *J. Phys. Chem. Solids* **61**, 585-592 (2000).

15. A. Ghosh, S.K. Bandyopadhyay, P. Sen, P. Barat and A.N. Basu, "Critical exponents of α-irradiated $Bi-2212$ superconductors in the paracoherence region", *Mod. Phys. Lett. B 12*, 829-837 (1998).

16. S.-C. Zhang, "A unified theory based on $SO(5)$ symmetry of superconductivity and antiferromagnetism", *Science* **75** 1089-1096 (1997); "The $SO(5)$ theory of high-T_c superconductivity", *Physica C* **282-287**, 265-268 (1997).

17. M. Pekala, E. Maka, D. Hu, V. Brabers, and M. Ausloos, "Mixed-state thermoelectric and thermomagnetic effects of a $Bi_2Sr_2CaCu_2O_{8+\delta}$ single crystal", *Phys. Rev. B* **52**, 7647-7655 (1995).

18. J.C. Soret, L. Ammor, B. Martinie, Ch. Goupil, V. Hardy, J. Provost, A. Ruyter and Ch. Simon, "Magnetoresistance in Bi-2212 single crystals", *Physica C* **220**, 242-248 (1994)

19. M. Ausloos and Ch. Laurent, "Thermodynamic Fluctuations in the Superconductor $Y_1Ba_2Cu_3O_{9-y}$. Evidence for two dimensional superconductivity", *Phys. Rev. B* **37**, 611-613 (1988).

20. P. Peyral, C. Lebeau, J. Rosenblatt, J.P. Burin, A. Raboutou, Q. Pena and C. Perrin, "Phénoménologie et caractérisation des supraconducteurs à haute T_c", *J. Phys. III France* 1, 1815-1821 (1991).

21. S.A. Sergeenkov, "On excess magnetoconductivity of a superconductive glass", *Z. Phys. B* **82**, 325-327 (1081).

22. M. Ausloos, Ch. Laurent, S. K. Patapis, S. M. Green, H. L. Luo and C. Politis, "Evidence for Anomalous Fluctuations in Superconducting $Bi_{1.75}Pb_{0.25}Ca_2Sr_2Cu_3O_{10}$", *Mod. Phys. Lett. B* **2**, 1319-1325 (1988).

23. M. Ausloos, P. Clippe and Ch. Laurent, "Homogeneous and Fractal Behavior of Superconducting Fluctuations in Electrical Resistivity of Granular Ceramics Superconductors", *Phys. Rev. B* **41**, 9506-9508 (1990).

24. D.Stauffer and A.Aharony, *Introduction to Percolation Theory*, (Taylor & Francis, London, 1992) 2nd printing

25. C.P. Burgess and C.A. Lütken, "$SO(5)$ invariance and effective field theory for high-T_c superconductors", *Phys. Rev. B* **57**, 8642-8655 (1998).

26. D.P. Arovas, A.J. Berlinsky, C. Kallin and S.-C. Zhang, "Superconducting Vortex with Antiferromagnetic Core", *Phys. Rev. Lett.* **79**, 2871-2874 (1997).

LIST OF PARTICIPANTS

Dr. AUSLOOS Marcel
SUPRAS, Institut de Physique B5
Université de Liège, Sart Tilman
B-4000 LIEGE, Belgium

Ms. BALDOVINO Sylvia
Inorganic Chemistry
University of Turin
Via Gaidano, 18
10137 TORINO, Italy

Dr. BARILO Sergei
Lab. Supercond. Materials Physics
Inst. Solid State & Semicond. Physics, ASB
17 P. Brovka Str.
MINSK 220072, Belarus

Pr. BENSALEM Mohamed
Lab. Physique des Materiaux
Fac des Sciences de Bizerte
BIZERTE, Tunisia

Dr. BLANK Dave H.A.
Low Temperature Div.
Department of Applied Physics
University of Twente
TWENTE, NL

Pr. BOULESTEIX Claude
Lab. MATOP
Fac. des Sciences St-Jerome
Av. Escadrille Normandie-Niemen
13397 MARSEILLE Cedex 20, France

Dr. CLOOTS Rudi
S.U.P.R.A.S., Inst. de Chimie, B6
Univ. de Liège - Sart Tilman
B-4000 LIEGE, Belgium

Pr. DABROWSKI Bogdan
Dept. of Physics
Northern Illinois University
FW 216
DEKALB, IL, 60115, USA

Dr. DAMAY Françoise
Experimental Solid State Group
Blackett Laboratory - Imperial College
Prince Consort Rd.
LONDON SW7 2BP, U.K.

Dr. de la FUENTE Xerman
Instituto de Ciencia de Materiales de Aragon
Centro Politecnico Superior
c/María de Luna, 3
E 50015 ZARAGOZA, Spain

Dr. DEMCHENKO V. A.
Institute of Thermoelectricity
Academy of Sciences of Ukraine
General Post Office Box 86
CHERNOVSTY, Ukraine

Pr. DEW-HUGUES David
Department of Engineering Science
Oxford University
Parks Road
OXFORD OX1 3PJ, U.K.

Dr. DIKO Pavel
Institute of Experimental Physics
Slovak Academy of Sciences
Watsonova, 47
04353 KOSICE, Slovakia

Pr. EMELCHENKO Gennadi
Crystal Growth
Institute of Solid State Physics - RAS
142432 CHERNOGOLOVKA, M.D., Russia

Dr. GABOVICH Alexander M.
Crystal Physics Department
Institute of Physics, NASU
Prospekt Nauki, 46
252022 KIEV, Ukraine

Pr. GILABERT Alain
LPMC
Université Nice Sophia Antipolis
Parc Valrose
06108 NICE Cedex 02, France

Pr. GOODENOUGH John B.
Texas Materials Institute
26th & SanJacinto / Bldg. ETC 9.102
Univ. of Texas at Austin
AUSTIN, TX 78712, U.S.A.

Dr. GRIGORESCU Cristiana
INOE 2000
P.O. Box MG-5
76 900 BUCHAREST, Romania

Dr. HOUSSA Michel
Silicon Process Technology Division
IMEC
Kapeldreef 75
B-3001 LEUVEN, Belgium

Dr. HURD Alan J.
New Materials Theory & Validation Dept.
Sandia National Labs.
ALBUQUERQUE, NM 87185-0333, U.S.A.

Dr. ILIESCU Brandusa
Natl. Institute for Materials Physics
P.O. Box MG-7
76 900 BUCHAREST, Romania

Dr. KLETOWSKI Zbigniew
Institute of Low Temp. & Structure Research
Polish Academy of Sciences
P.O.Box 1410
50-950 WROCLAW, Poland

Dr. KOPIA-ZASTAWA Agnes
Faculty of Metallurgy
Sciences of Materials
30. Al. Mickiewicza
PL- 30-059 KRAKOW Poland

Pr. KRESIN Vladimir
Lawrence Berkeley Laboratory
University of California
1 Cyclotron Rd. / MS 62/203
BERKELEY, CA 94720, U.S.A.

Pr. KURMAEV Ernst Z.
Inst. of Metal Physics
Russian Acad. Sci., Ural Div.
S. Kovalevskaya 18, GSP-170
620219 YEKATERINBURG, Russia

Dr. LI Qi
Department of Physics
Penn. State Univ.
UNIVERSITY PARK, PA 16802, U.S.A.

Dr. LIU Ru-Shi
Dept. of Chemistry
National Taiwan University
N1, Roosevelt Road, Sect. 4
TAIPEI, Taiwiai, ROC

Dr. MAJCHROWSKI Andrzej
Institute of Applied Physics
Military University of Technology
Kaliskiego, 2
00-908 WARSAW, Poland

Pr. MANEVAL Jean-Paul
LPMC
Ecole Normale Superieure
24, rue Lhomond
75005 PARIS, France

Pr. MOLENDA Janina
Department of Solid State Physics
University of Mining and Metallurgy
30, Al. Mickiewicza
30 059 KRAKOW, Poland

Dr. NAKAMAE Sawako
Physique de l'Etat Condensé
DRECAM-DSM
CEA Saclay- Bat 771
F-91101 GIF/YVETTE, France

Pr. NEDKOV Ivan
Institute of Electronics
Bulgarian Academy of Sciences
72, Tzarigradko Chaussee
1784 SOFIA, Bulgaria

Dr. NIKONOV S. Yu.
Iofee Physical Technical Institue
R A S
St PETERSBURGH , 194021 Russia

Dr. NIKULOV Alexey V.
Inst. Microelectr. Technol. & High Pur. Mater.
Russian Academy of Sciences
142432 CHERNOGOLOVKA, M.D., Russia

Dr. ODIER Philippe
Lab. de Cristallographie - CNRS
25 Av des Martyrs - BP 166
38042 GRENOBLE Cedex 09, France

Dr. PAVUNA Davor
Physics Dept.
IPA - EPFL
Swiss Federal Institute of Technology
CH-1015 LAUSANNE, Switzerland

Dr. PEKALA Marek
Dpt. of Chemistry
Univ. Warsaw
Al. Zwirki i Wigury 101
PL-02-089 WARSAW, Poland

Dr. PLECENIK Andrej
Institute of Electrical Engineering
S A S
Dubravska cesta 9
84239 BRATISLAVA, Slovakia

Pr. ROBBES Didier
GREYC & CRISMAT
ISMRa
6, Bd du Marechal Juin
14050 CAEN Cedex, France

Dr. SCHMITZ Georg J.
Materials Sciences
ACCESS e.V.
Intzestr. 5
D-52072 AACHEN, Germany

Pr. SEIDEL Paul
Institut für Festkörperphysik
Friedrich-Schiller-Universität Jena
Helmholtzweg, 5
D 07743 JENA, Germany

Dr. SERGEENKOV Sergei
Bogoliubov Lab. Theoretical Physics
Joint Institute for Nuclear Research
Joliot-Curie, 6
DUBNA, Moscow Region, Russia

Dr. SILVA Enrico
Dipartimento di Fisica "E.Amaldi"
Universita' di Roma Tre
Via della Vasca Navale 84
00146 ROMA, Italy

Dr. TAKADA Jun
Dept. of Applied Chemistry
Okayama University
3-1-1 Tsushima-naka
OKAYAMA 700-8530, Japan

Dr. TAMPIERI Anna
IRTEC
CNR
Via Granarolo, 64
48018 FAENZA (RA), Italy

Dr. TIKARE Veena
Materials Theory & Computation
Sandia National Laboratories
PO Box: MS 1411
ALBUQUERQUE, NM 87185-1411, U.S.A.

Pr. URUSHADZE Givi
Tbilisi State University
Chavchavadze Ave. 75, Korpus 9, Kvartira 41
380062 TBILISI, Georgia`

Dr. VACQUIER Gilbert
LPCM
Univ. de Provence
Centre St-Charles - 3, Pl. V. Hugo
13331 MARSEILLE Cedex 3, France

Dr. VANNIER Rose-Noëlle
L.C.P.C.S.
ENSCL - BP 108
59652 VILLENEUVE D'ASCQ Cedex, France

Ms. VERTRUYEN Benedicte
S.U.P.R.A.S.
Inst. de Chimie, B6
Univ. de Liège - Sart Tilman
B-4000 LIEGE, Belgium

Dr. VILLEGIER Jean-Claude
Laboratoire de Cryophysique
SPSMS
CEA-Grenoble, 17 rue des Martyrs
38054 GRENOBLE Cedex 9, France

Dr. WELCH David O.
Materials & Chemical Sci. Div.
Dept. Applied Sci.
Brookhaven National Lab.
UPTON, NY 11973, U.S.A.

Pr. ZIPPER Elzbieta
Institute of Physics
University of Silesia
ul. Uniwersytecka, 4
40-007 KATOWICE, Poland